THE MILITARY SYMBOLS JOURNAL AND PLANNER #1

A PRIMER FOR SOLDIERS, MARINES AND MILITARY ENTHUSIASTS

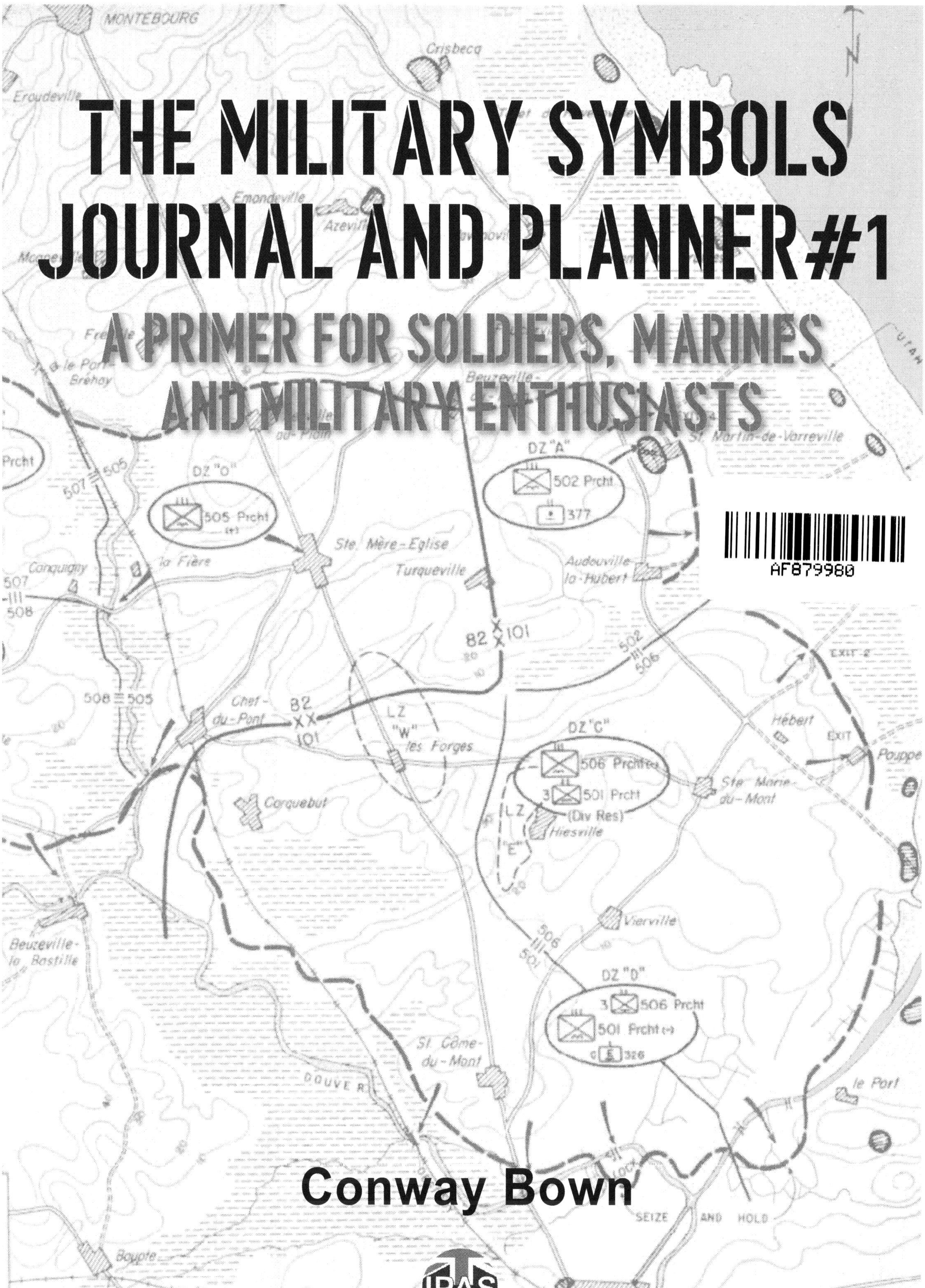

Conway Bown

IPAS

WHY THIS BOOK?

This book has been created for anyone that has a need for, or an interest in, being able to understand and render military symbols. Soldiers and Marines, commissioned, warrants, non-commissioned and enlisted, will use these symbols as a part of briefing and planning, and must be able to demonstrate knowledge of them during their promotion courses and when delivering briefings to superiors and subordinates or during educational sessions.

Members of headquarters staffs, operations staffs and intelligence units will also use these symbols often, either electronically by way of electronic battle maps, or the 'old school' method of pen and paper.

Military historians or those giving lectures and presentations of historical events will also find a knowledge of military symbols extremely useful, as can be seen by the front cover of this journal/planner where the landing zones of the US airborne divisions have been depicted showing drop zones, landing zones, formation and unit boundaries and significant targets.

Some weapons systems used in modern warfare may not be used today as warfare has changed, but these military symbols allow for them to be depicted using the rules of construction making it useful for historians, authors and illustrators. Even if the weapon system may be unknown to a modern soldier, the symbol will tell you what its capability was. A thorough understanding of the symbols helps to reduce barriers to communication and gets a message across clearly and succinctly.

HOW TO USE THIS BOOK.

This journal/planner assumes a year of use, and so each two-page spread contains five sections representing five work days per week, as well as a weekly lesson on military symbology. But there's no reason to stick to this pattern. Use as many day-sections as necessary... or skip a few if there's nothing to write down. So long as each section is dated appropriately, then it will be simple to find the day you need. There are some extra pages to cover extra military symbology.

By reading and practicing each week's lesson on military symbols, then by the end of the book the reader will have an expert-level knowledge of all the military symbols he or she will need. Or if you need to learn it quickly, then it also provides that opportunity for 'cramming' the knowledge.

I hope you find this journal/planner useful and informative.

Conway Bown

ISBN 978-0-6452466-3-6

www.ipas.com.au

This book is dedicated to the men and women who, by their own volition,
put on a uniform and put themselves in harm's way,
as professionals or volunteers, in the service of their fellow citizens,
and to the families and loved ones that support them.

To all those that have died in the service of their country and community,
may they rest in peace and honour.

	JAN	FEB	MAR	APR	MAY	JUN	
1							1
2							2
3							3
4							4
5							5
6							6
7							7
8							8
9							9
10							10
11							11
12							12
13							13
14							14
15							15
16							16
17							17
18							18
19							19
20							20
21							21
22							22
23							23
24							24
25							25
26							26
27							27
28							28
29							29
30							30
31							31

	1	2	3	4	5	6	7	8	9	10	11	12	13	14	15	16	17	18	19	20	21	22	23	24	25	26	27	28	29	30	31
JUL																															
AUG																															
SEP																															
OCT																															
NOV																															
DEC																															
	1	2	3	4	5	6	7	8	9	10	11	12	13	14	15	16	17	18	19	20	21	22	23	24	25	26	27	28	29	30	31

UNDERSTANDING MILITARY SYMBOLS

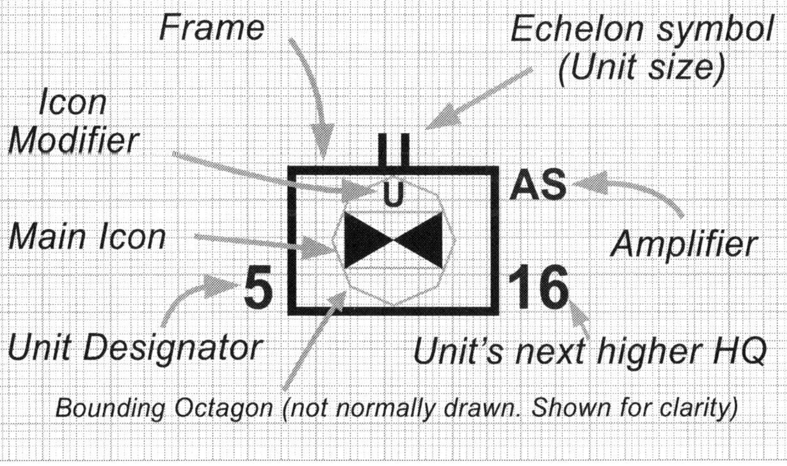

A military symbol carries information about a unit, installation, capability, piece of equipment or weapon system. They can also depict events. But the most common land symbols used predominantly by Army and Marines will be what we're reviewing in this journal/ planner.

On the left are the main components of a military symbol. This one depicts the Australian 5th Aviation Regiment of the 16th Aviation Brigade operating utility helicopters. How do I know? Read on and find out!

2

SYMBOL SHAPES

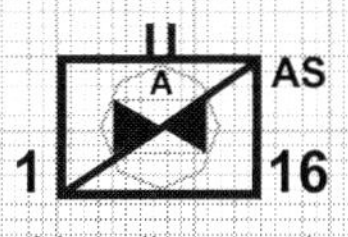

A friendly attack helicopter regiment

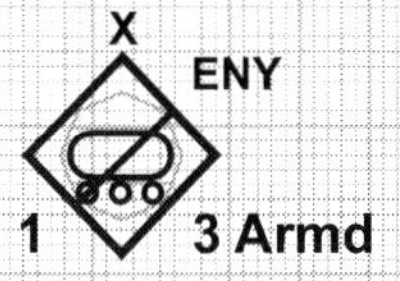

An enemy armoured recce brigade

A neutral search and rescue airplane

A light machine-gun

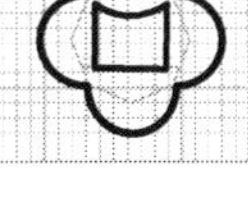

A utility vehicle of unknown affiliation

Military symbols are shown in different colours to indicate affiliation. On the rear cover of this book are the various shapes and colours used: blue/cyan for friendly, red for hostile, green for neutral, yellow for unknown and purple for civilian. These work well on electronic battlemaps or when drawn on a map using coloured marker pens.

Alternatively, when using just black and white, the shape and accompanying text provide the required information.

Rectangles are friendly units/installations, diamonds are hostile, squares are neutral entities, clover leaves are unknowns and circles are items of equipment or weapons.

The Bounding Octagon

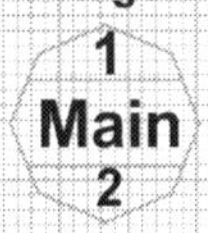

Inside each shape fits a bounding octagon which is not drawn on the symbol, but which is shown here as it helps to create the symbol. The bounding octagon has three sections. The main section carries the primary icon and sections 1 and 2 provide modifying icons to the main icon.

3

MODIFIERS AND AMPLIFIERS

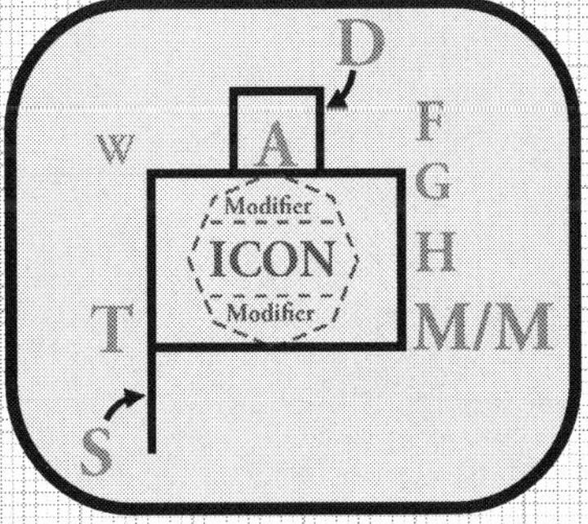

Military symbols have modifying or amplifying symbols attached that provide more information about the symbol. The icon and modifiers go inside the frame and amplifiers that provide extra 'amplifying' information go outside the frame. There are several, but the following are the most common with their relative positions shown at left:

ICON - the symbol denoting what the entity is or its type - Main section.
MODIFIERS - Top section: special capabilities / characteristics. Bottom section: mobility, size.
AMPLIFIERS:
A - Echelon symbol (size symbol, eg platoon, squad, brigade, etc)
D - Task force indicator (when applicable)
T - Unit designator (usually the unit's number or letter)
M - Higher HQ (can also be shown as T / M / M etc on either side of the icon)
W - Date time group/s indicating when the object occupied that location
S - HQ amplifying symbol (flag mast added, when applicable)
F - Reinforced or reduced (denoted by (+) or (-))
G - Comments (eg ENY - enemy, NATO country code, etc)
H - Comments (eg equipment type / common name of equipment)

SIZE (ECHELON) AMPLIFIER

Military symbology uses unit size amplifiers to explain what size a unit is or what area of responsibility is controlled by a unit. It is important to know these symbols but not every country has the same sized units or formations as others.

ø	A fire team or crew
●	A squad (US) or section (C'th)
●●	A section
●●●	A platoon or troop or section
I	A company or battery or squadron or troop
II	A battalion or squadron or regiment
III	A regiment (US - usually consisting of several battalions)
X	A brigade
XX	A division
XXX	A corps
XXXX	An army

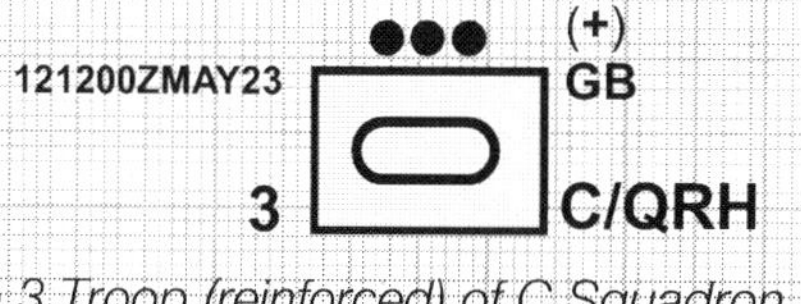

No 3 Troop (reinforced) of C Squadron, Queen's Royal Hussars, British Army, occupying a position at 1200Z on 12 May 2023.

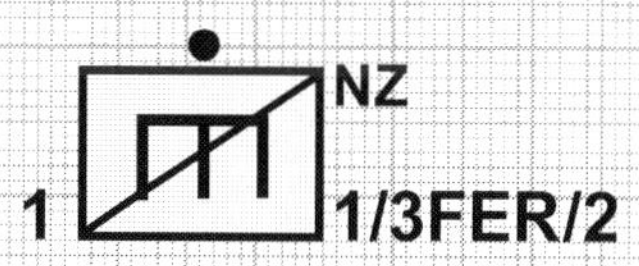

No 1 Section (in a reconnaissance role) from 1 Troop, 3rd Field and Emergency Response Squadron, 2nd Engineer Regiment (New Zealand Army)

5

HOW TO READ A MILITARY SYMBOL

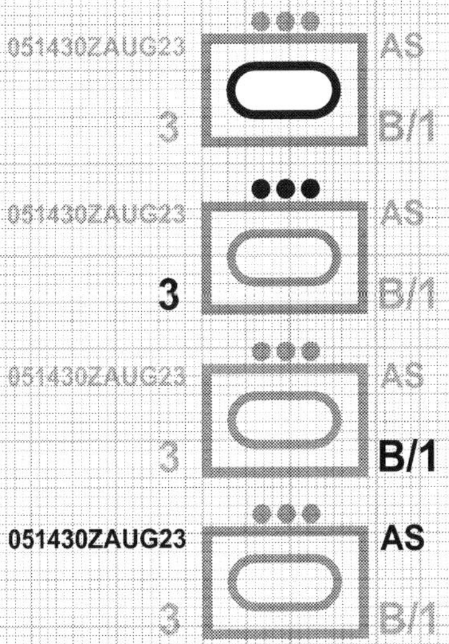

1. First we interpret the icon and any amplifying marks. In this case the icon is the oval shape signifying tracked armour... so this is an armoured unit.

2. Next, we read the unit designator and echelon symbol. The designator for this unit is '3' and the echelon (unit size) symbol is the group of three dots of a platoon, or troop depending on the type of capability. In this case, being armoured, it is a troop. No 3 Troop

3. After determining it is a troop of armour, namely '3' Troop, we can read the higher echelon/s in its chain of command. It is read left to right with each character being the next higher HQ. B/1 means B Squadron, 1st Armoured Regiment. In some instances a hyphen is used when a unit belongs to a regiment, eg 2-506 = 2nd Battalion of the 506th Regiment. This is an exception to the standard unit structure and naming conventions. It reflects historical structuring and is normally kept for heritage reasons.

4. Lastly, we read any amplifiers. AS stands for Australia and 051430ZAUG23 is a date time group... so this symbol indicates that 3 Troop of B Squadron, 1st Armoured Regiment, Australian Army is/was occupying a location at 1430hrs on 05 Aug 2023

INFANTRY AND ENGINEERS

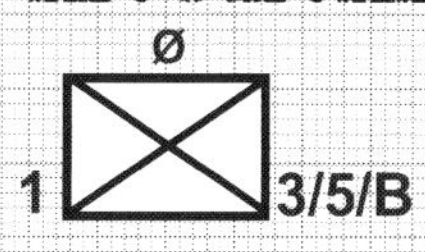

The icon for infantry contains two diagonal slashes representing the old backpack straps worn by the infantry as shown in this image of British soldiers of the 18th Century.

The echelon symbol is a slashed zero which represents an infantry fireteam, specifically Fireteam 1, 3rd Squad, 5 Platoon, B Company

The letter 'E' represents field engineers, and it being turned on its side is reminiscent of a truss bridge that engineers would build to cross a river or a tank ditch.

This icon shows 2nd Platoon, A Company, 2nd Engineer Battalion, US Army.

7

CAVALRY (RECCE) AND SIGNALS

C  2

2  A

The diagonal slash of cavalry/reconnaissance evokes the Sam Browne belt as worn across the torso of this Canadian Mountie. The Sam Browne was named after its creator who had lost his arm in battle and needed a strap to hold his sword's scabbard so he could draw it. It came to represent soldiers who used swords or sabres, such as the cavalry, whose primary role is reconnaissance. This unit is C Squadron, 2nd Cavalry Regiment in tracked armoured vehicles.

The lightning flash of electricity and the electro-magnetic spectrum has been used to symbolise the role of signals which had its origins in semaphore and phone lines and then radio. Today, in the age of satellite communications, it still signifies the lightning quick transfer of messages and information.

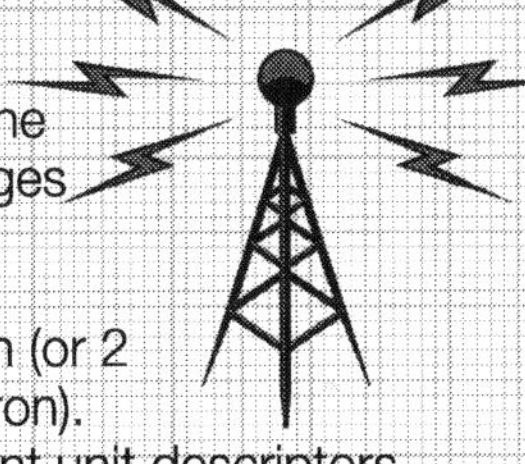

The icon above depicts 2 Platoon (or 2 Troop), A Company (or A Squadron). Different countries will use different unit descriptors.

8

AVIATION AND MAINTENANCE

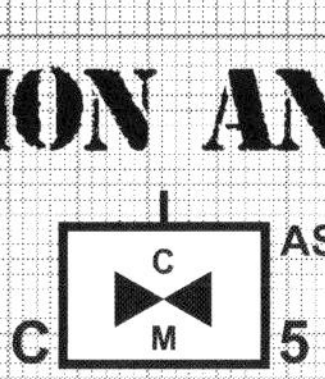

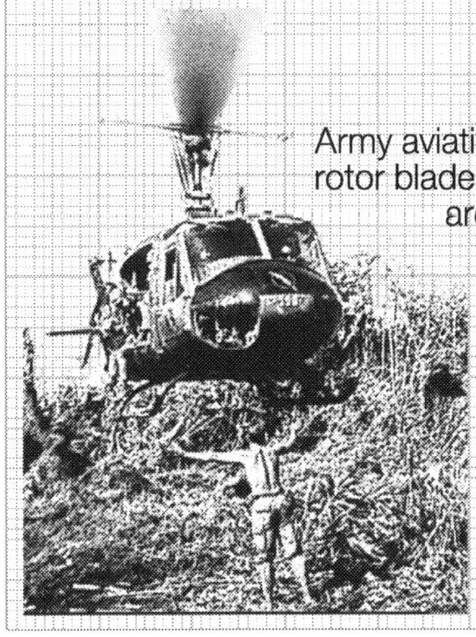

Army aviation is denoted by the blurred rotor blades of a helicopter. The blades are fully coloured as if they are camouflaged. This differentiates them from a fixed-wing aircraft symbol which is a propellor that is coloured if it is Army and uncoloured if it is not.

The icon shows C squadron of 5th Aviation Regiment flying medium (M) cargo (C) helicopters.

Within Commonwealth armies, maintenance functions on major equipment are normally carried out by electrical and mechanical engineers. The icon for EME is a spanner but the spanner icon can also be a modifier for other capabilities. The icon above is a squadron of EME for a transport regiment. The picture and icon at right shows a Royal Canadian EME recovery team from B Squadron, Lord Strathcona's Horse (Royal Canadians) recovering a damaged tank in Korea.

9

TRANSPORT AND HEALTH

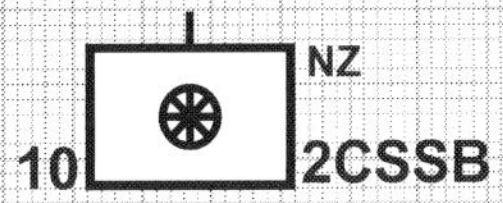

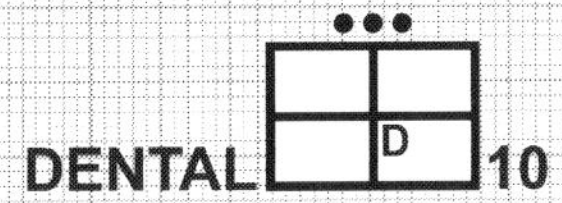

A cart's wheel is the symbol for transport units and evokes the horse and carts of supply trains of armies through the centuries.

The icon above shows the 10th Transport Company of the 2nd Combat Service Support Battalion of the New Zealand Army.

The health capability consists of many constituent specialities: surgical, dental, psychological... even veterinarian.

The symbol for health is the cross of the Red Cross. If a particular speciality needs to be indicated it is written off-centre in the bottom right. The icon above is dental section of the 10th Field Hospital.

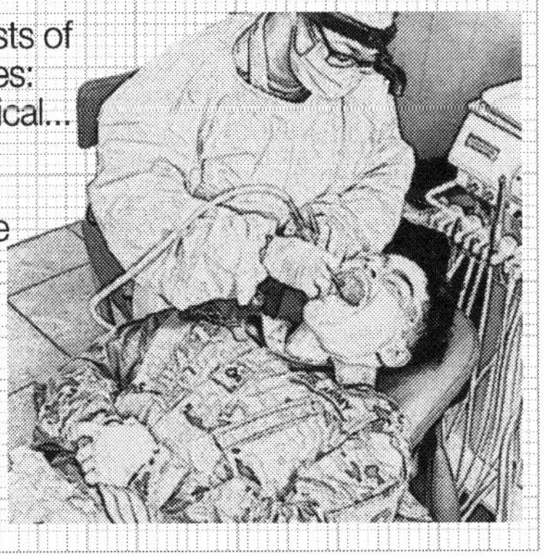

10

AIR DEFENCE AND AMPHIBIOUS

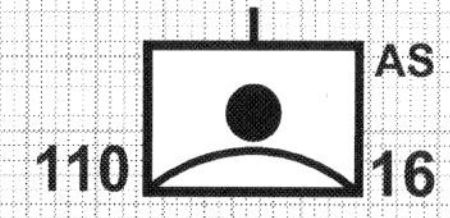

Some capabilities have specific items of equipment which might be integrated into the military icon. The symbol for anything air defence is the arc of a missile through the air, which will be seen in other symbols. The icon above is 110 Battery of 16 Air Land Regiment, Royal Regiment of Australian Artillery.

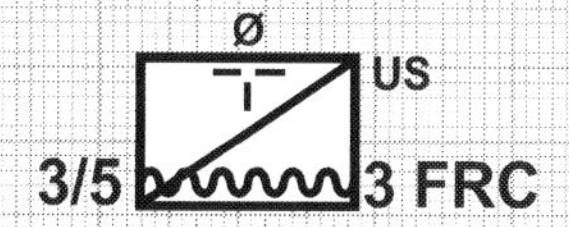

Amphibious capabilities are denoted by the wavy lines of the water, usually on the bottom of the frame, but can also be centred. This icon shows an amphibious reconnaissance team as denoted by the slashed zero (team/crew) and the diagonal slash (recce). The three dashes in the top part of the frame depict a sight marking, which denotes that this is a sniper team, specifically SSP 3 of 5 Platoon, 3rd Force Recon Company, USMC.

11

AIRBORNE AND HOSPITALS

E  2-506/101

8055 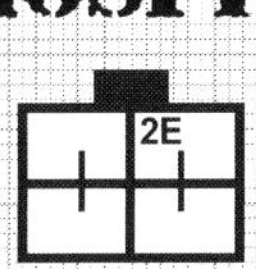 MOBILE

Of course, learning military symbols is useful for the study of military history also. ***Band of Brothers*** is the story of E Company (or 'Easy Company' using the old phonetic alphabet) of the 2nd Battalion, 506th Airborne Infantry Regiment of the 101st Airborne Division. The airborne infantry symbol is the gull wings modifier. As the term 'Regiment' this context, this is addressed in weeks 4 and 5.

The TV series *M*A*S*H* was based on the real life 8055th Mobile Army Surgical Hospital during the Korean War. The symbol above denotes the 8055th as an installation, depicted by the black rectangle, but with the amplifier 'Mobile'. A MASH was about company size, so an alternative could be to use the symbol for a company. The icon is the medical cross, but with two small bars to create the letter 'H'. The 2E denotes it is a Role 2 (Enhanced) medical facility which a Korean War MASH would equate to.

12

MOBILITY OR CAPABILITY

The bottom section of the icon, either within, or in line with the section 2 of the octagon, is the location where often is placed modifier 2 which details the type of capability or the type of mobility of the unit or item of equipment. We saw in the previous icon for airborne that the gull wings were placed in the lower third. Likewise with the recon sniper team in the amphibious task icon the week before last where the waves of the water were placed in the bottom third. Both these examples show how the unit moves, or its special capability for movement or operation. Modifier 2 can also be used to categorise the unit's or equipment's size, such as light, medium or heavy.

On the right are some common modifier symbols used to denote mobility or special capability or size.

A brigade of light infantry

A troop of wheeled armour (eg LAV/Stryker) in a reconnaissance role

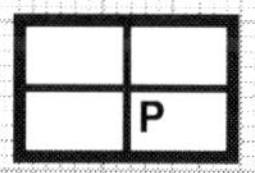

A psychological medical team

A company of water transport (eg landing craft)

An aviation maintenance recovery team

13

HEADQUARTERS

A Headquarters is where the commander and his staff will operate to command a unit or formation. When the command and control is from that location, then the symbol has a vertical line, called the command staff (where staff means a flagpole) attached to the symbol. The command staff must be at least the same height as the octagon. In order to precisely indicate the HQ on a map, an offset line can be drawn from the bottom of the command staff to the exact location of the HQ. Offset lines on other symbols can come from another boundary side of the frame, but must not be made to look like a command staff. See the two examples at right of how offsets are used.

Depending on the size, or in other circumstances where the HQ functions need to be separated, a horizontal bar in the top third of the frame represents a headquarters element, or a function carried out at a headquarters. In this example, a transport platoon is part of 2 Division HQ, but is not the element that commands 2 Div, so a uses a horizontal bar—not a command staff—to denote it is part of the HQ, but not the command element. In the US Army, the Headquarters and Headquarters Company (HHC) unit and name describes this separation well.

The location of a battalion HQ denoted by the HQ staff modifier and the offset indicator pointing the the precise location of the HQ.

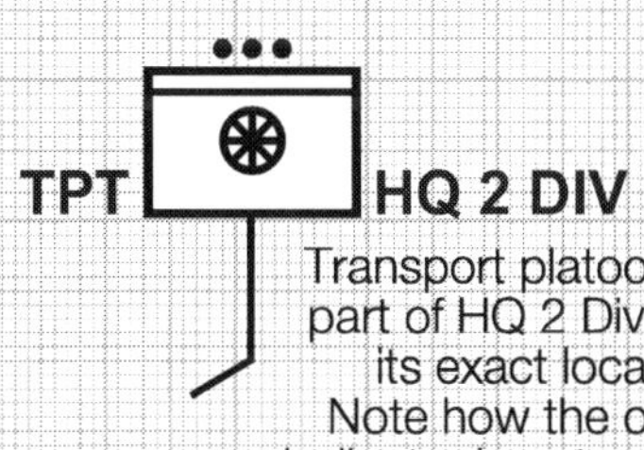

Transport platoon as part of HQ 2 Div and its exact location. Note how the offset indicator is not part of the HQ staff.

14

OCCUPYING AN AREA

If a unit is to be shown occupying an area, such as a defensive position, then an almost complete line surrounding the area occupied is drawn on the map with the echelon symbol placed in the gap. If the enemy is to be shown on the map or inferred, then the gap and the echelon symbol are shown on the other side of the shape facing away from the enemy.

The figure at right shows 2 Platoon, A Company from a battalion of light infantry (letter 'L' as a modifier in the symbol) occupying an area with an enemy unit to its north.

The enemy position is also indicated using the occupied area shape but with the abbreviation for enemy - ENY - shown on the sides of shape. The diamond unit symbol likewise denotes enemy which, in this case, is an infantry unit of unknown size or identity.

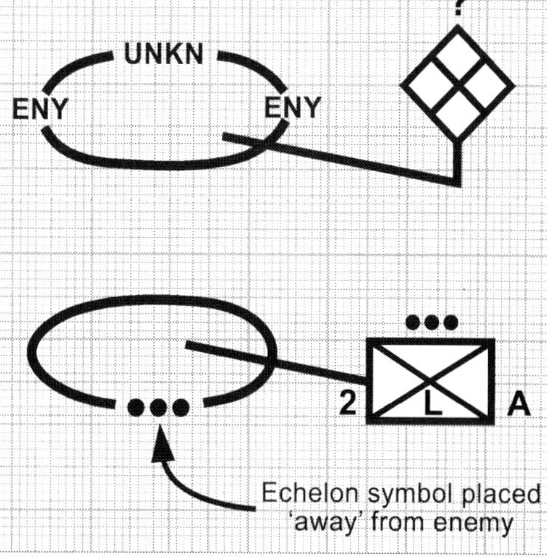

15

LOGISTICS AND SELF-PROPELLED ARTILLERY

If a horizontal line above the icon means the headquarters, then underpinning everything is logistics (supply), and so it is represented by a horizontal line underpinning any icon. The symbol shows a logistics platoon that provides refuelling and petrols, oils and lubricants (POL) for aircraft.

A cannonball and tank tracks represent self-propelled artillery. The sIG 33 was a German self-propelled artillery weapon system based on the chassis of a Panzer tank. This sIG belonged to the 17th sIG battalion assigned to the 22nd Panzer Division in Stalingrad and is shown within a 'hostile' shaped symbol.

16

AMPHIBIOUS ARMOUR & AVIATION RECCE

3 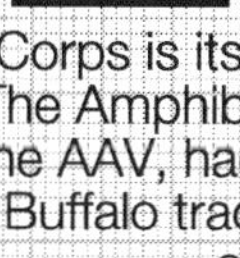 1 MARDIV

A unique feature of the US Marine Corps is its amphibious armoured transport capability. The Amphibious Assault Vehicle, the AAV, had its genesis with the Buffalo tracked landing craft of WWII.

This symbol shows the 3rd Assault Amphibian Battalion of the 1st Marine Division.

29 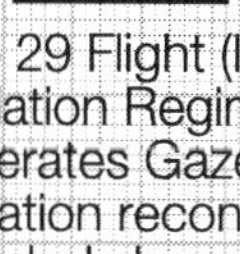 GB 665

No 29 Flight (Independent) of No 665 Squadron, 5th Aviation Regiment, Army Air Corps of the British Army operates Gazelle helicopters. Its primary role is aviation reconnaissance. The military symbol shows all those aspects: the blurred rotor blades and recce diagonal slash) but also includes the common name of the equipment type below the country and above the higher unit echelon, 665 Sqn.

17

NATO CLASSES OF SUPPLY

There are slight differences in depicting the classes of supply between NATO and US miltary systems. NATO has five classes whereas the US has ten, and not all the symbology is consistent. The NATO military symbols at right show logistics supply (denoted by the horizontal line in the frame) and the different types supplied items per class.

Below is a way to remember what each class stands for.

CLASS I: Food & water are the first items required for survival.
CLASS II: All the equipment a soldier needs to do his job.
CLASS III. You've got food and vehicles, but you need fuel.
CLASS IV. Now that you have everything, you must prepare your position to defend it.
CLASS V. You have food, weapons, vehicles fuel, a constructed base... time to get ammo and get to work.

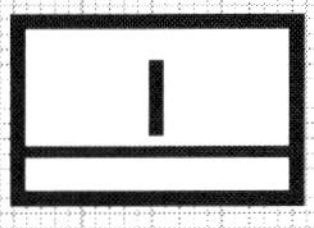

Class I: Items of subsistence consumed by personnel or animals (rations, etc)

Class II: Supplies as per organisational tables (eg clothing, weapons, tools, vehicles).

Class III: Petrols, oils, lubricants. (Aviation POL is Class IIIa)

Class IV: Construction and defensive stores and extra issues of Class II.

Class V: Ammunition, explosives and chemical agents.

18

US CLASSES OF SUPPLY PT 1

The US military system is extremely capable when it comes to supply, so there are five extra classes denoted in military symbols. This is a result of the differences in doctrine between the US Army and US Marines such that now the list is a joint list resulting in ten discrete classes. The first five are the same as for NATO, but the symbology is different, as shown at right. Below are some ways to remember them and their icons.

Read the NATO definitions as a starting point to remember the classes of supply.

CLASS I. The first thing you need to survive, food and water, which you gather during the day and eat when night falls.
CLASS II. The Quartermaster/supply sergeant has the keys to the stores where you find uniforms, combat gear, etc.
CLASS III. POL has three letters and POL uses a funnel.
CLASS IV. Defensive stores, such as what Engineers would use.
CLASS V. Same as NATO.

Class I: Items of subsistence consumed by personnel or animals (rations, etc)

Class II: Supplies as per organisational tables (eg clothing, weapons, tools, vehicles).

Class III: Petrols, oils, lubricants. (Aviation POL is Class IIIa)

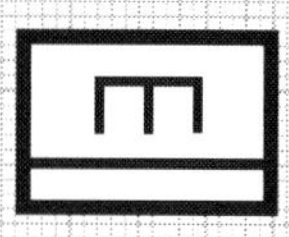

Class IV: Construction and defensive stores and extra issues of Class II.

Class V: Ammunition, explosives and chemical agents.

19

US CLASSES OF SUPPLY PT 2

Continuing on from last week are the US Classes of Supply. Classes 1 through 5 are the same as NATO even if the symbols are different. This week we look at classes 6 through 10 and their symbols and some ways to remember them.

CLASS VI. Personal items not supplied by the logistics system. You know... the small stuff you'd buy before going into battle, like chocolates and tabasco sauce for the rat packs/MREs.
CLASS VII. Major items, what they call 'end to end', and the icon reflect that. This includes weapon systems, vehicles, tanks... you know... the big stuff you need for the battle!
CLASS VIII. After the battle is over, time to patch up the wounded. You'll need some medical supplies. That's what CLASS VIII is.
CLASS IX. The equipment needs to be repaired, also! So CLASS IX is repair parts and tooling.
CLASS X. The battle over, time to help the locals. All the Civil Affairs supplies are CLASS X.

Class VI: Personal items and discretionary items. Alcohol, magazines, candy.

Class VII: Major items, weapon systems, tanks, aircraft, etc

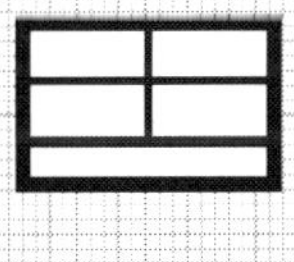

Class VIII: Medical supplies and specialised medical equipment.

Class IX: Repair parts and tooling and workshops.

Class X: Civil Affairs stores and supplies.

20

MILITARY INTELLIGENCE AND MILITARY POLICE

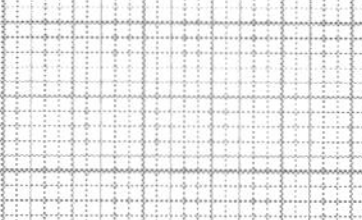

Some military capabiiities are denoted only by their initials, such as Military Intelligence. The military symbol above shows the 25th Military Intelligence Company of the 2nd Military Intelligence Battalion of the British Army.

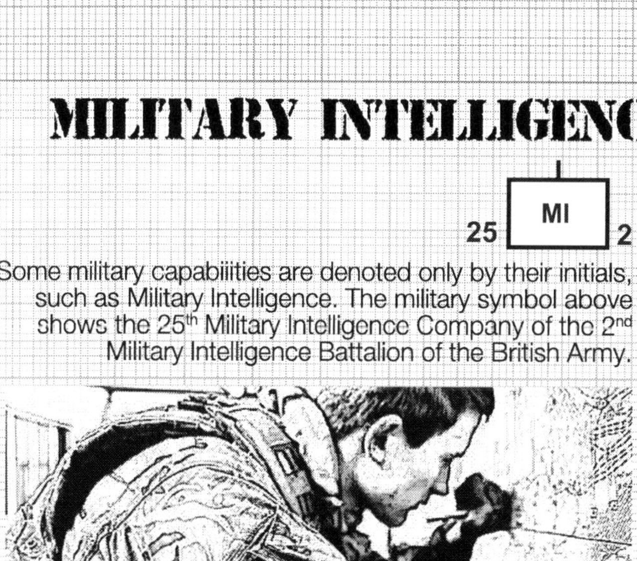

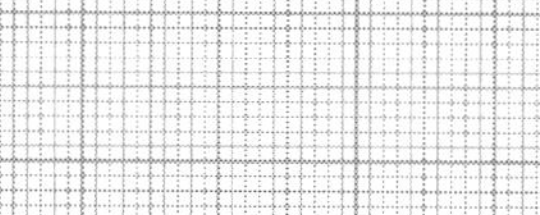

Like Military Intelligence, the Military Police symbol is merely the letters MP within the frame. Any modifiers would be added as necessary.

In this case, the MP platoon indicated has a special capability, namely it's a canine platoon, so the symbol has been amplified with both the 'K9' unit designator and modified with the word 'DOG' as a special capability in the section 1 modifier. So it is canine platoon of the 716th MP Battalion.

21

MORE MODIFIERS AND AMPLIFIERS

A vertical line within the frame on the left side indicates a main gun system. In the symbols at right, the top symbol shows an armoured vehicle with three circles denoting that it is wheeled and not tracked. Three wheels indicates it has cross-country capability while two wheels indicates limited or no cross-country capability. This symbol has text to the right where the common name of the main equipment can be written, so in this case it is a troop of LAV 25s, No 4 Troop specifically, with 25mm cannon. No 4 Troop is from B Squadron, 2nd Cavalry Regiment.

Below that symbol is the symbol indicating C Squadron of the 2nd/14th Light Horse Regiment, also operating wheeled armoured vehicles, but in this case they have infantry aboard, so this is mounted infantry.

The curly bracket is used to indicate on a map that the two sub-units are grouped for a purpose. We might assume the squadron of mounted infantry are being escorted by LAV 25s to provide protection and firepower.

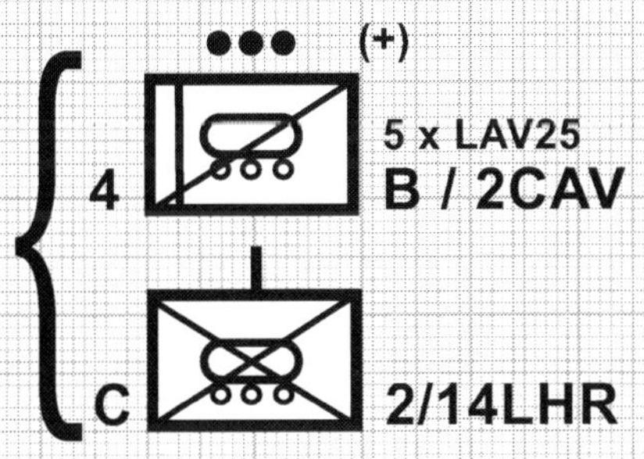

22

BOUNDARY LINES

To coordinate operations between units that are situated next to each other (adjoining), boundary lines are indicated on a map showing the extent of their tactical areas of responsibility (TAORs) or areas of operations (AOs).

Usually, if an element or sub-unit of one unit wishes to enter the TAOR or AO of an another unit, this must be coordinated so as to avoid misidentifying friendly troops as being enemy troops.

The boundary lines have an echelon symbol incorporated representing the highest level of the unit and AO concerned, and then identifying designators are placed within the respective TAOR/AO to denote which unit/formation has responsibility for that TAOR/AO.

In the diagram at right, a British battalion in the north shares a boundary with an Australian battalion which shares its southern boundary with a New Zealand battalion. The NZ battalion TAOR is further broken down into company TAOR s. Each boundary has an echelon symbol.

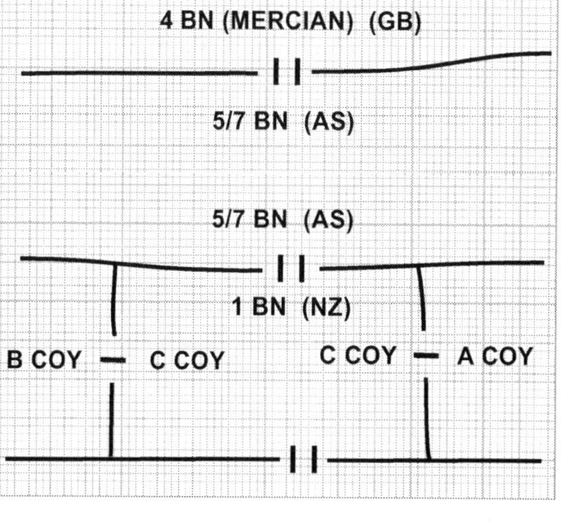

23

FIXED-WING AVIATION

225  OV-1D / SLAR 223 Avn Bn

Where the symbol for an Army or Marine helicopter are the triangle-shaped blurred rotor blades, the symbol for a fixed-wing aircraft is the propellor. If it is an Army or Marine aircraft the propellor is filled in (as if it has camouflage paint on it), but if it is Air Force, the propellor is not filled in. The symbol above denotes 225th Aviation Company of the 223rd Aviation Battalion (Combat) operating OV-1D Mohawks during the Vietnam War. The Mowhawks were equipped with Side Looking Airborne Radar denoted by the radar modifier above the icon and the SLAR abbreviation as an amplifier. SLAR was able to pick up moving objects out to 45 kilometres. These aircraft were used to detect enemy trucks on the Ho Chi Minh Trail in Laos without violating airspace.

UAS

10  GB 47

Ø VMU3  US MAG 24

Unmanned Aerial Systems in the British and Australian Armies were initially operated by the corps of artillery and, at time of writing, still are.

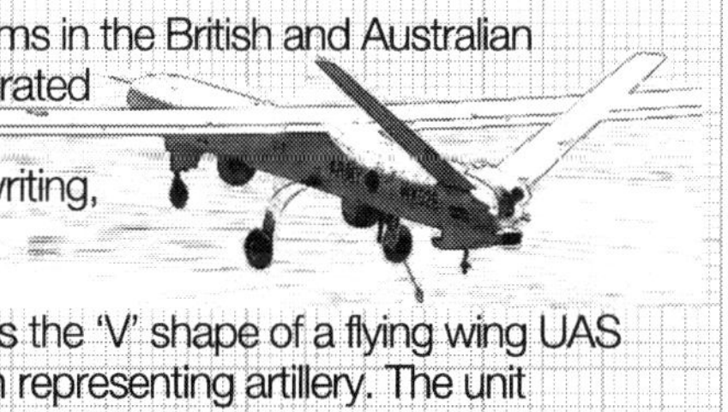

The modifier for a UAS is the 'V' shape of a flying wing UAS with the cannonball icon representing artillery. The unit symbolised on the left is No 10 Battery, 47 Regiment, Royal Artillery, while the crew on the right is a recovery team (recovery modifier is a stylised net below the icon) of USMC UAS squadron VMU3, part of Marine Aviation Group 24. Note the difference in the main icons, but the same role.

24

SIGNALS

Technology through the ages has seen the way we communicate on the battlefield change markedly. With direct SATCOM these days the need for a retransmission site on the highest mountain in the AO is all but gone... but not quite. Retrans, as it is known, is still needed, but devices like Aerostat airships or UAS able to loiter often replace the two-man teams sitting alone, on a mountain with a radio waiting for a re-supply of rations and water.

APOE/APOD

An air point of entry or air point of departure (APOE/APOD) is/are locations where personnel or equipment are expected to arrive or depart. They are usually established airports or airfields capable of multiple aircraft arrivals and departures of different sizes, but that is operation dependent. The symbol for such a location is a combination of runways and the symbol for transportation, the cartwheel. They are usually thought of as installations.

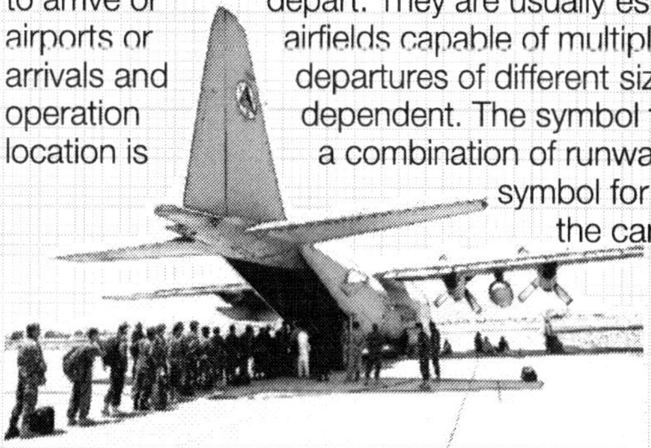

25

A point symbol indicates an entity or an event that was, is or will be located at a specific point. Its text fields are similar to those of other military symbols. If it is a future event, then the symbol will use a dashed outline (explained later).

The key concern is that it provides information that is relevant for the persons likely to view it without cluttering up a map or diagram.

The image immediately at right explains where the modifiers and amplifiers go and what information they provide.

At far right are point symbols showing an aviation refuel location - 'Fido' - supplied by 2 Squadron of 3rd Aviation Regiment, able to refuel CH-47s and smaller. The other example is the start point (SP) for an activity under the responsibility of 6th Battalion which is effective at/from 1200Z on 20 July 2023.

POINT SYMBOLS

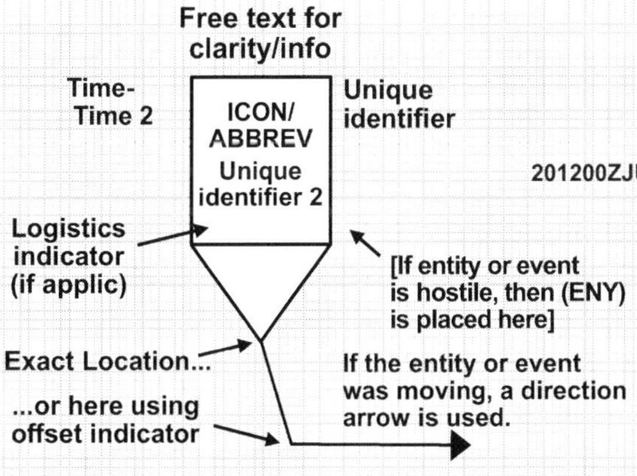

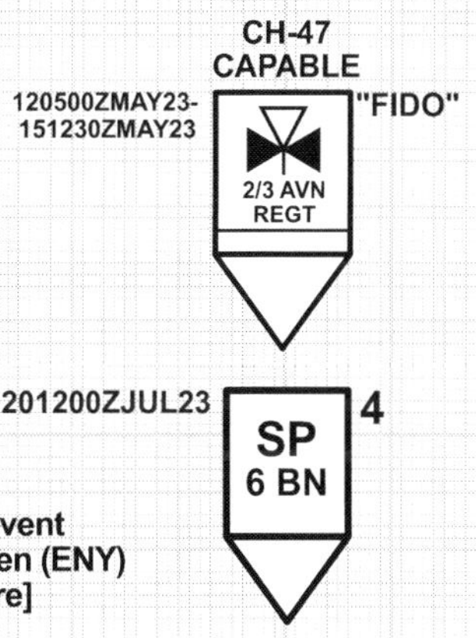

26

POINT SYMBOLS - RESUPPLY

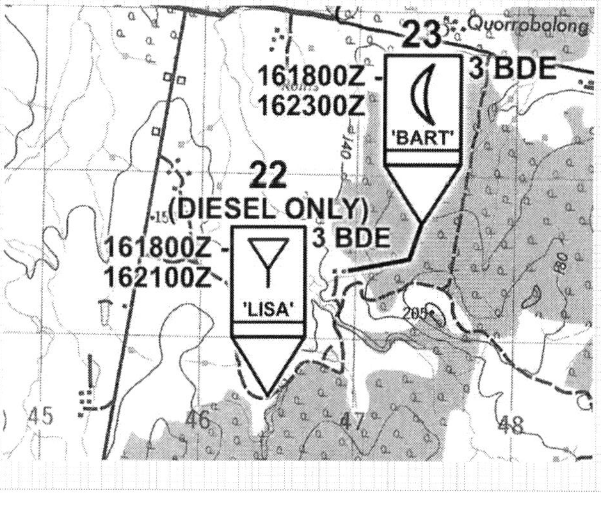

Carrying on from last week's lesson on point graphics. On the map at left two resupply points have been indicated: Bart and Lisa.

Bart is a Class 1 resupply (rations, etc) as denoted by the 'moon' symbol (the US symbol for Class 1) and the horizontal line beneath the code name. (See mil symbols lesson from 7 weeks ago). It will be open from 1800 hrs on the 16th and will close at 2300Z hrs that day. It is being supplied by 3rd Brigade and it has an identifier number '23'. It has been offset from its location to avoid cluttering the map, so an offset indicator indicates the location of the resup point.

The other resup point, 'Lisa', is for refuel but has only diesel fuel. It is also supplied by 3rd Brigade and will be open from 1800Z and will close at 2100Z on the 16th. It is located on the bend of the track.

27

MAINTENANCE AND RECOVERY

The role of the modifier sections of the bounding octagon allow for an icon to signify different things.

For example, when a small spanner icon is in the upper section it represents a maintenance function <u>related to the capability</u> depicted by the military symbol.

When a large spanner icon is in the main icon section (the middle), it indicates that the unit or installation's <u>primary</u> function is maintenance.

When a symbol is in the bottom section, is signifies a <u>special capability, size, movement or environment</u>, so a spanner shown here is for recovery.

The symbols at right show how the placement of a modifier changes the function/capability of the represented unit or installation.

A maintenance and logistics installation

A maintenance troop or platoon of an armoured unit.

An aviation unit's recovery section or squad.

28

PLANNED, FUTURE, SUSPECTED

Using a dashed line to draw symbols and boundaries indicates that the friendly or neutral entity or activity is planned to occur or be located at a specified time or place. If it represents enemy, then a dashed line indicates 'suspected' or 'forecast'. After all, you can never be too certain when you deal with enemy actions, can you? You can only go by best guess or by intelligence reports.

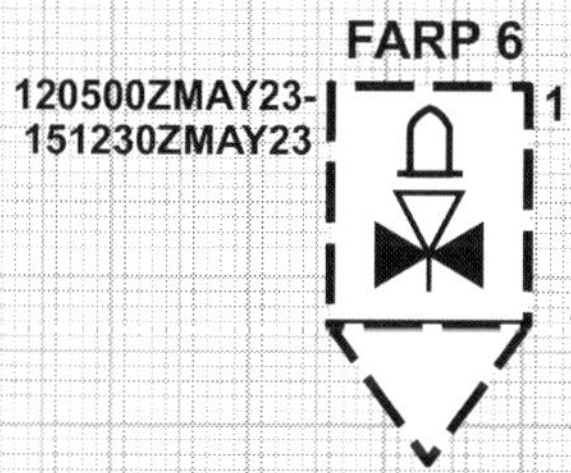

The planned location and open/close timings of a Forward Arming and Refuelling Point for aircraft. (Instead of icons it could also just use the word "FARP").

Suspected location of an enemy self-propelled artillery regiment HQ. (Note the HQ Staff symbol denoted by the vertical line.)

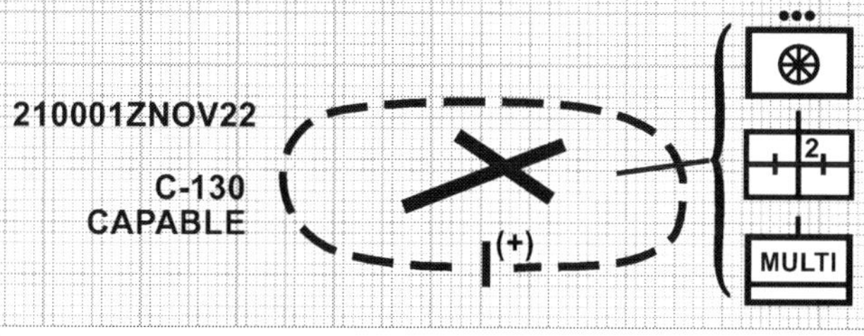

The planned occupation of an airfield zone from 0001Z on 21 Nov 2022 of a two company-sized grouping of a transport platoon, a Role 2 hospital and a multiple class logistics node. C-130 Hercules can land there.

29

AREAS AND LOCATIONS

An area currently occupied by No 3 Platoon

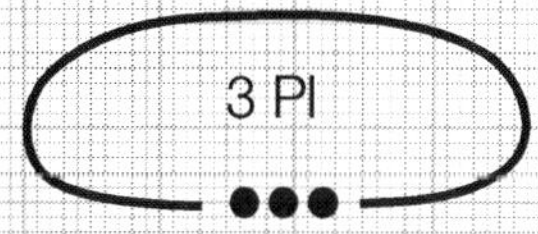

Observation Post / Outpost

Drop Zone for E Co. 2-506 PIR

An area to be occupied by C Company at 0600Z on 4 Apr, 2023

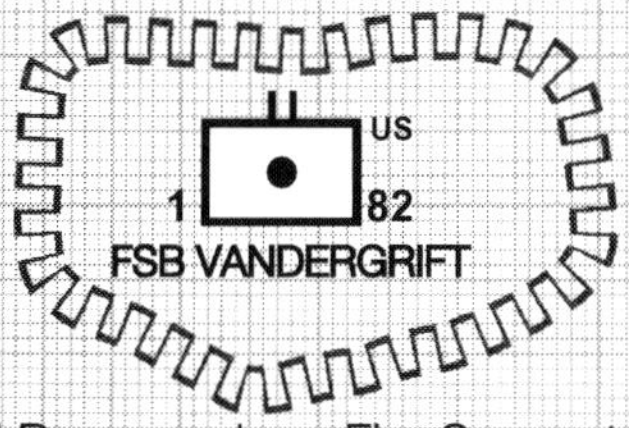

A Fortified Base, such as Fire Support Base Vandergrift in South Vietnam, where 1st Bn, 82nd Field Artillery Regiment was located.

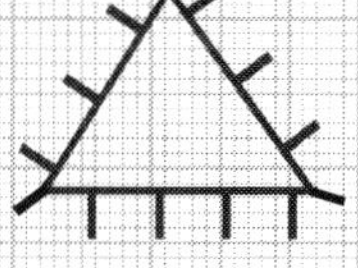

A Combat Outpost. The lines pointing outwards are reminiscent of rifles emanating from the walls of a fort, ready to defend it.

30

WEAPON SYSTEMS AND VEHICLE SYMBOLS

Military symbols can denote entities, units, installations and even events. After denoting units, probably the most common type of symbol is that of weapons systems and vehicles.

On static (ie paper) maps, they can indicate a location where a weapon or a vehicle was spotted. On an electronic battlemap, the icons may represent a vehicle moving in real time and being tracked by satellite or drone.

There are some design features of a weapons symbol that once learned will make it simple to interpret them.

Let's start...

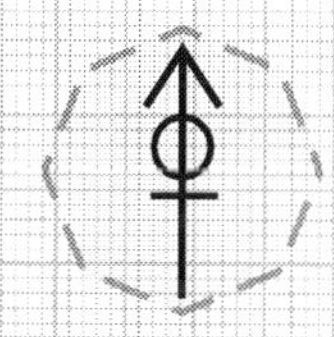

A light grenade launcher

The weapon system or the vehicle symbol must be drawn such that it remains within the bounding octagon. The octagon itself is not drawn but is merely a guide.

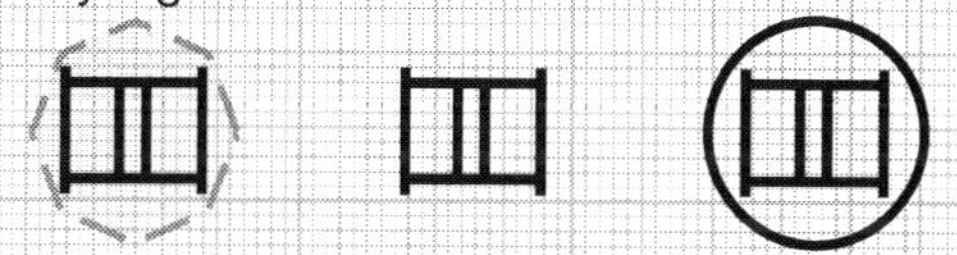

A medium tank shown within its bounding octagon to show dimensions. The symbol can be drawn by itself (middle) or within a circle (right). The circle merely indicates that it is a piece of equipment, but can be discarded to keep a diagram clean and uncluttered. If using the circle, a dashed circle indicates enemy when drawn in black and white, otherwise it is shown in red.

31

WEAPON SYSTEMS - SYMBOL COMPONENTS

There are two key components of a military symbol that represent a land-based weapon system: a central vertical line, and two horizontal lines. A weapon system symbol with a central vertical line ususally sends, launches or discharges a projectile of some type, whereas a symbol with at least two horizontal lines or curves is a vehicle of some description.

Think of the central vertical line as being the barrel or launcher, and think of the horizontal lines as being the front and back or top and bottom of a vehicle. After those key components are rendered, their special capabilities can be created by adding shapes or other lines to make unique symbols and icons that represent classes of weapons or vehicles.

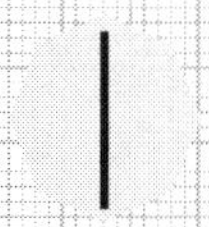

A barrel, launcher or rail. A basic requirement for a weapon system and the basis for military symbols representing weapon systems.

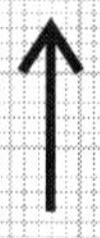

A rifle. Its base symbol.

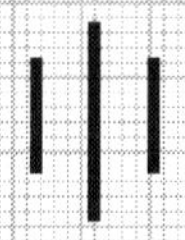

An artillery system. Think of the two vertical lines being the wheels.

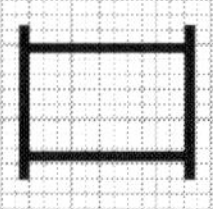

A tank.

An armoured personnel carrier.

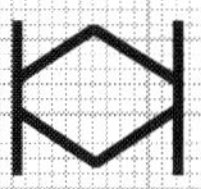

An armoured fighting vehicle.

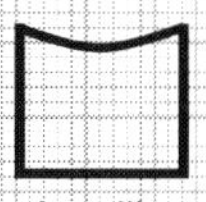

A utility vehicle (ie not a vehicle specifically designed for combat).

32

SYMBOL CHARACTERISTICS - SIZE INDICATORS

By adding modifying lines to a symbol, the relative size can be shown. Weapons have horizontal size modifiers and vehicles have vertical size modifiers, as a general rule.

A howitzer, size not shown

A light howitzer

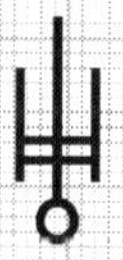
A medium howitzer

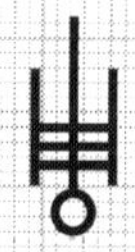
A heavy howitzer

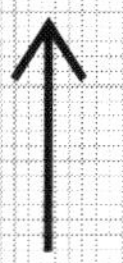
A rifle, size not shown

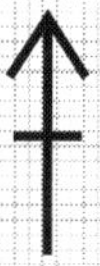
A single-shot rifle

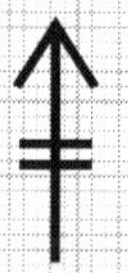
A semi-automatic rifle

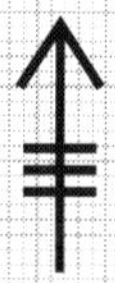
An automatic rifle

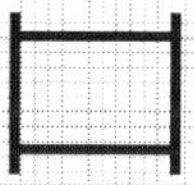
A tank, size not shown

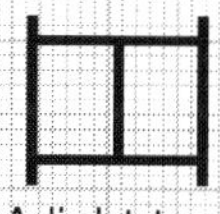
A light tank

Each weapon or vehicle has a basic shape. When the size or capability is known, the modifier then provides the viewer with size and/or capability.

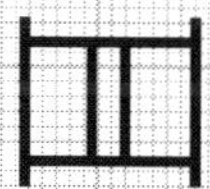
A medium tank

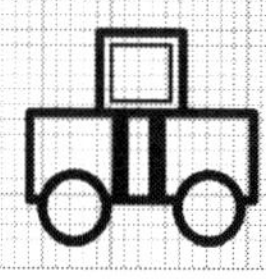
A mid-sized sedan (civilian)

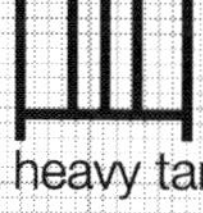
A heavy tank

33

SYMBOL CHARACTERISTICS - MORTARS & GRENADE LAUNCHERS

On the left is a mortar and pestle. Below left is a 17th Century mortar. It was named after the mortar and pestle because of its shape. Below is a pile of mortar shot. The military symbol for a mortar is a mortar shot at the base of its launcher with the arrowhead sybolising that you should know by now, indicates it is a medium system, say 61mm mortar or so. the weapon is firing a projectile. The two horizontal lines,

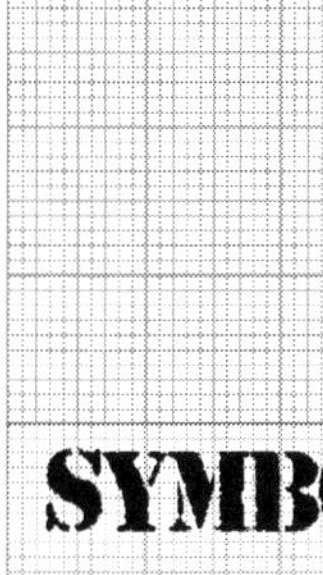

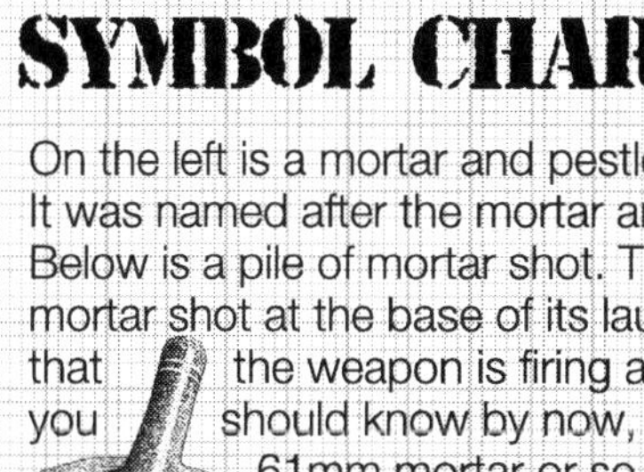

Mortar shot in the bottom of the mortar

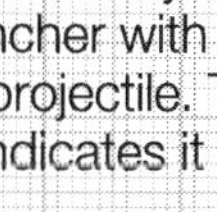

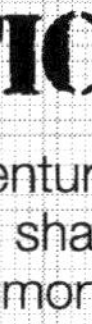

A medium mortar system

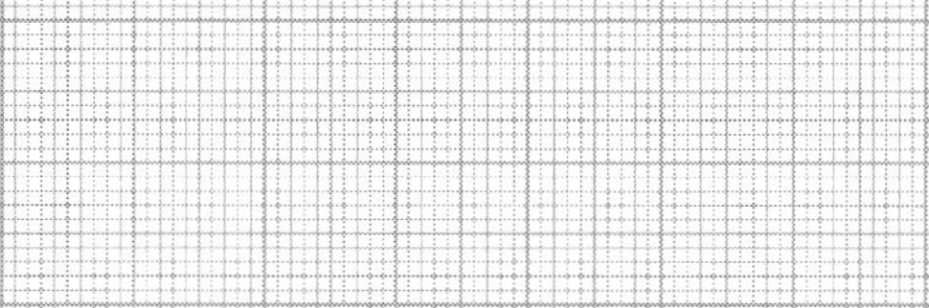

Grenade situated further along the launcher or barrel

A light grenade launcher

The symbol for a grenade launcher is similar to a mortar, but in most cases the grenade launcher has the grenade held further up the launching device or tube, such as a grenade launcher assembly for a rifle.

34

SYMBOL CHARACTERISTICS - ROCKETS AND MISSILES

As a basic rule, missiles are usually guided and rockets are not, or have minimal guidance.

Many rockets are used in the anti-tank role and so often have a pointed nose for armour piercing capabilities. Missiles on the other hand, usually have some sort of sensor in the nose such as TV or infra-red, and so its nose is often rounded.

These are the tell tales between rocket and missile systems in military symbols.

The rounded head of a missile, such as this Javelin, is reflected in the rounded head of the military symbol. The vertical line represents the launcher and the horizontal line shows that it is a 'light' system. The inverted 'V' is the symbol for anti-tank. So this is a light anti-tank missile system.

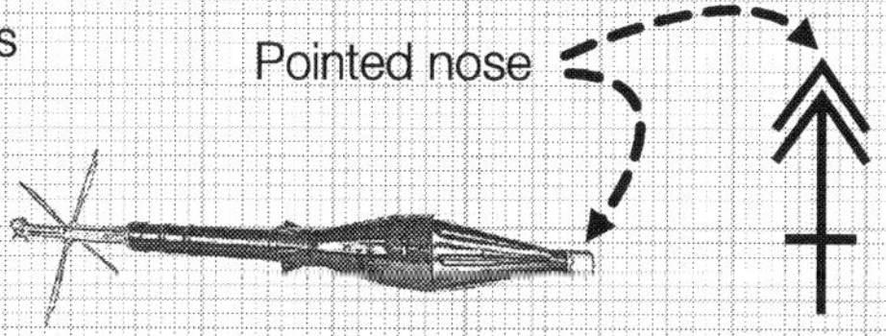

The pointed head of a rocket projectile is reflected in the military symbol. A good way of remembering this symbol is that one arrowhead signifies a projectile is discharged, and the second arrowhead represents the pointed nose of a rocket assisted projectile such as this rocket propelled grenade from an RPG-7. The military symbol represents a light, single-projectile, rocket launcher system.

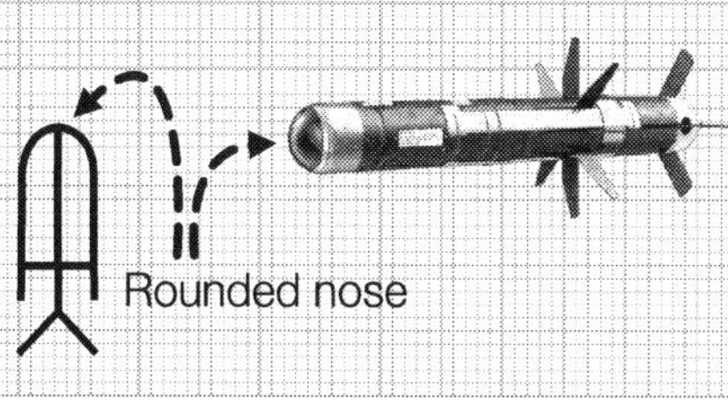

35

SYMBOL CHARACTERISTICS - AIR DEFENCE

As we've seen, the placement of vertical or horizontal lines will symbolise if the icon represents a weapon and its size, or a vehicle's size. This week we're looking at weapons pertaining to air defence.

See the image of the missile tracing an arc through the air? In military icons, anything with an arc in the icon means it has something to do with air defence such as anti-aircraft missiles or guns.

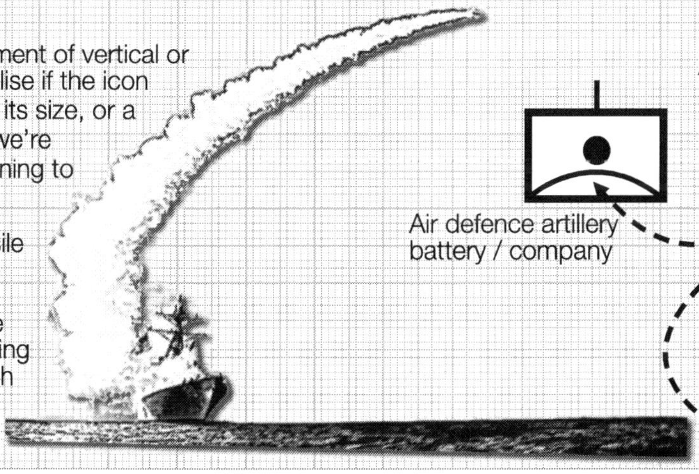

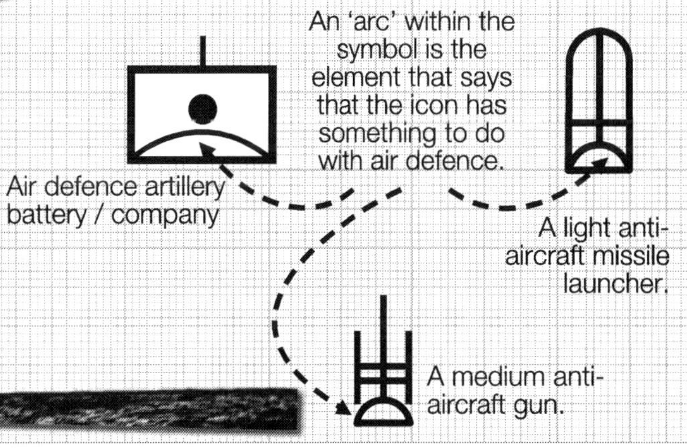

36

SYMBOL CHARACTERISTICS - ANTI-ARMOUR

Anti-armour ammunition can be HESH (High Explosive Squash Head), Discarding Sabot, HEAT or others. In almost all cases, the nose of the projectile will have a distinctive point, some even needle like, such as seen below. This feature makes it easy to remember the symbol characteristic for anti-tank as a point, or an inverted 'V', in the same way that an arc is anti-aircraft.

Anti-armour rounds have a distinctive point... so does the unique characteristic of anti-armour military symbols.

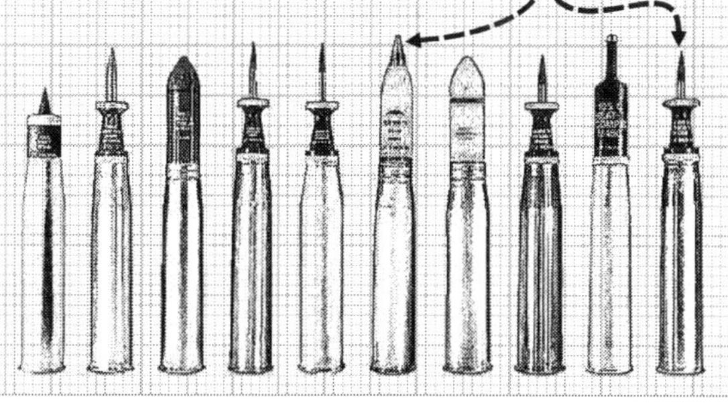

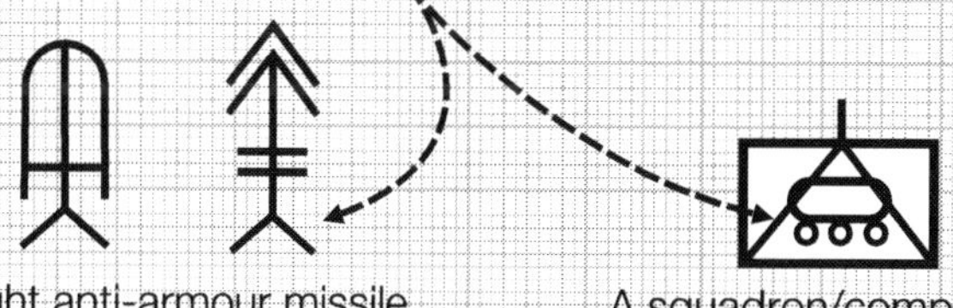

A light anti-armour missile system on the left and a medium anti-armour rocket system on the right

A squadron/company of anti-armour wheeled armoured vehicles such as Strykers or LAVs

37

SYMBOL CHARACTERISTICS - ARTILLERY

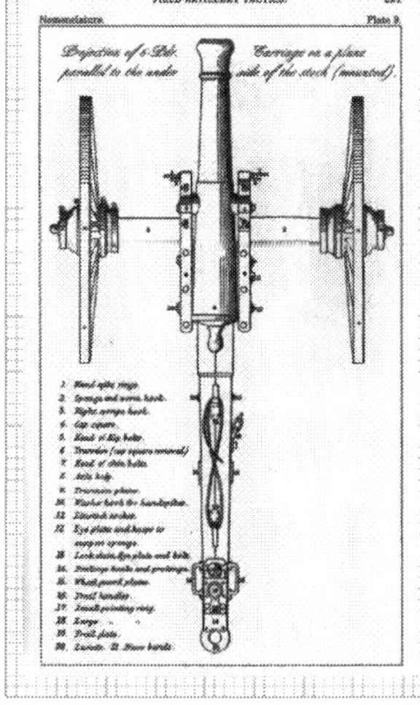

Artillery systems are usually denoted by the vertical line of the barrel but also two smaller vertical lines that could represent the wheels of the gun. Modifiers then tell the viewer what special characteristics the gun has.

The basic shape to represent an artillery cannon. See how it replicates an old cannon such as this Civil War era 6-pounder at left.

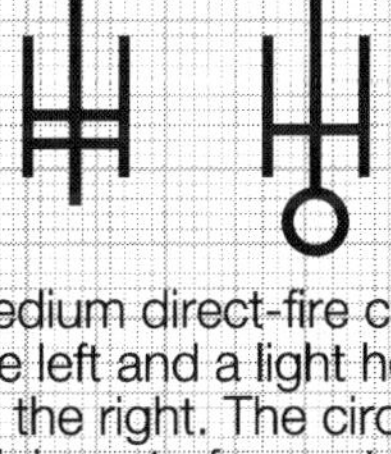

A medium direct-fire cannon on the left and a light howitzer on the right. The circle is reminiscent of a mortar and they both share the same high, arcing trajectory of the projectile.

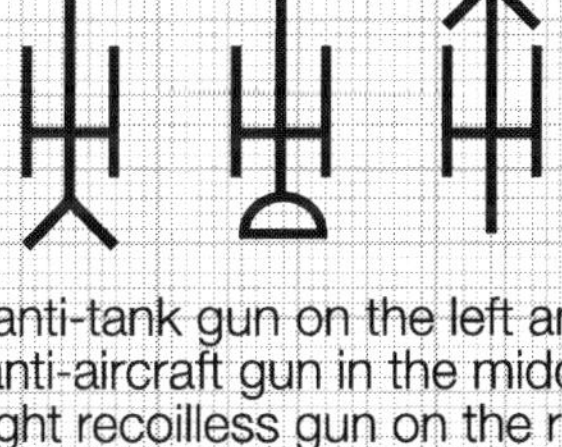

A light anti-tank gun on the left and a light anti-aircraft gun in the middle and a light recoilless gun on the right. Recoilless guns are usually rifled so an easy way to remember this one is the rifle symbol with the wheels of an artillery piece.

38

SYMBOL CHARACTERISTICS - MACHINE GUNS & OTHER MAN-PORTABLE WEAPONS

The symbol for machine guns is the same as for a rifle, however it has a horizontal line at the base. Think of it as something that stabilises the symbol just as a machinegun will have a bipod, tripod or pintle mount to stabilise it. The non-lethal weapons show the vertical line of the launcher or barrel, but a horizontal line at the top to signify that the projectile is not designed to kill.

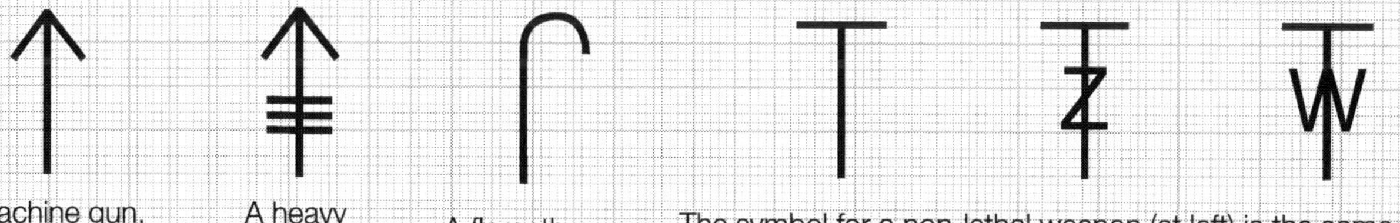

A machine gun, size not shown

A heavy machine gun

A flamethrower

The symbol for a non-lethal weapon (at left) is the same as the symbol for the task verb 'block' (discussed later). This is usually used by police forces. The middle symbol is for a Taser and on the right is a water cannon.

30

SYMBOL CHARACTERISTICS - MISSILES AND ROCKETS

In this second entry on missiles and rockets, we look at surface-to-surface weapons and how they are depicted. Surface-to-surface guided weapons have a flat base on the icon. Multiple launch rockets have a double arrowhead.

The MGM-140 Army Medium Tactical Missile System

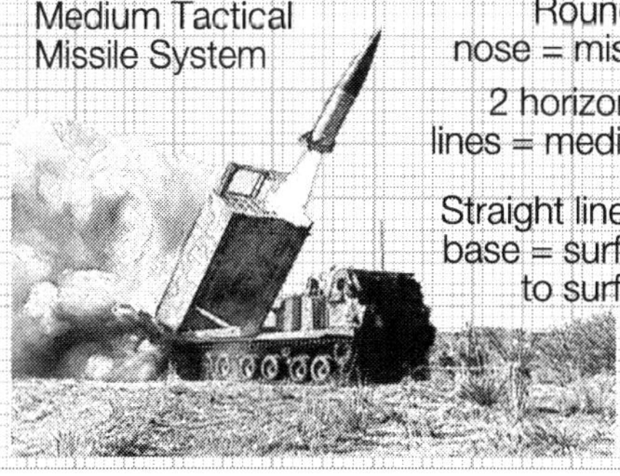

Vertical line = launcher/barrel

Rounded nose = missile

2 horizontal lines = medium

Straight line for base = surface to surface

Rocket symbol

Artillery weapon system, therefore able to fire mulltiple projectiles

1x horizontal line = light

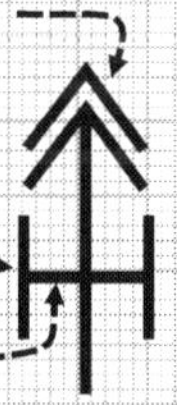

The HIMARS, HI Mobility Artillery Rocket System is considered to be a light rocket system and extremely mobile, able to be airlifted and driven into a theatre of operations at short notice.

40

SYMBOL CHARACTERISTICS - ARMOURED VEHICLES

Vehicles for military use, be they armoured (or armored if you're American), or unarmoured, will be depicted using an enclosed symbol. And just like the symbol for an artillery cannon, think of the two vertical lines as being like the tracks or wheels, and the rest of the shape depicting the type of vehicle. If it is necessary to distinguish the fact that the vehicle is wheeled, then small circles can denote that it is wheeled. For example, a mix of tracked and wheeled fighting vehicles.

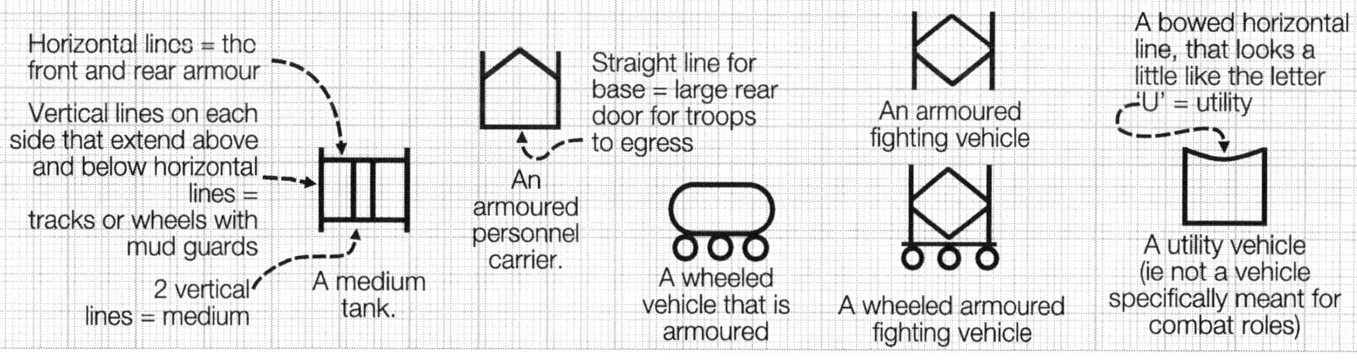

41

SYMBOL CHARACTERISTICS - ARMOURED VEHICLES

Specialised vehicles have modifiers added to their symbols to distinguish them and their roles, as shown below.

The symbol for an armoured recovery vehicle, such as this Australian M113 'Fitter' and RAEME crew changing the powerpack on an Australian Armoured Corps Centurion tank during the Vietnam War.

The symbol for an M548 tracked carrier as used by the 'Five Eyes' nations. This one is a Canadian carrier used for logistic resupply and so the symbol has the flat line in the lower third to denote logistics.

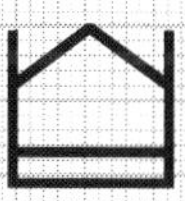

An APC in an ambulance role.

42

SYMBOL CHARACTERISTICS - UTILITY VEHICLES

The concept with military symbols is that they should be as uniform as possible, however provided that the correct components are used, and the viewer can interpret them, then the item can be inferred even if it is not 100% accurate.

The symbol for a wrecker/tow truck is a hook. The size is indicated by the number of crossed lines. This symbol shows a medium armoured, wheeled cross-country wrecker. In this case, the item has been enclosed in a circle, however the circle can be omitted if necessary to keep a map uncluttered.

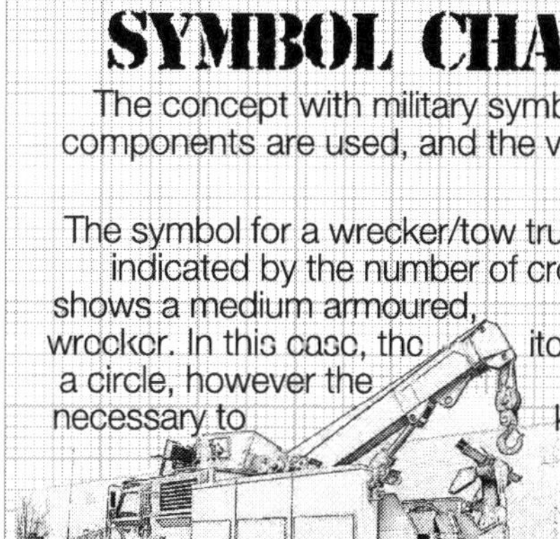

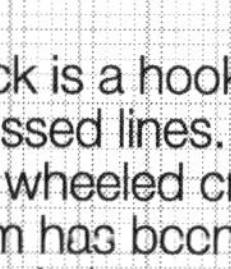

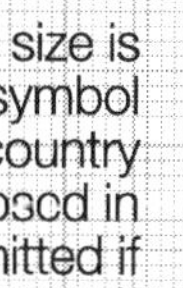

An ambulance in a utility vehicle.

Both symbols below denote a utility vehicle with a refuel role for aircraft. The image is of a WWII tanker used by the 8th Air Force stationed in Britain. The left symbol depicts a utility vehicle used for aircraft refuelling. The right symbol has more specific information: a wheeled utility aviation gasolene refuelling vehicle with limited cross-country capability.

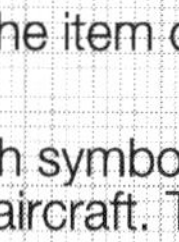

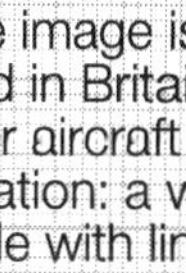

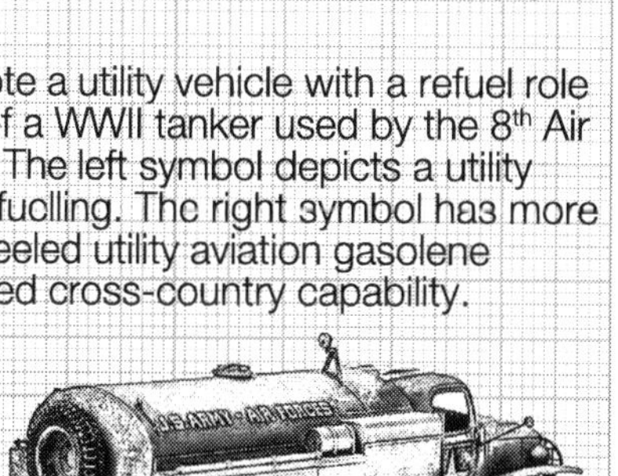

SYMBOL CHARACTERISTICS - OTHER VEHICLES

Engineering and transport vehicles will feature prominently in combat support and combat service support tasks. Using the rules of military symbology discussed over the last 43 spreads we can infer what each vehicle's function or capability is. In some cases, free text is part of the symbol, such as the 'Bus' below.

Engineer vehicles

A backhoe

A wheeled engineer vehicle

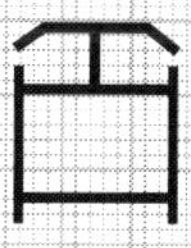
An earthmover on a tank chassis

Transport vehicles

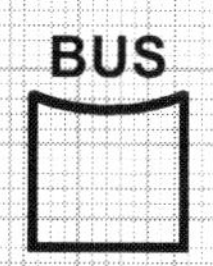

A bus

A medium tow truck

A semi-trailer and truck (medium)

An amphibious utility wheeled vehicle

44

COMMAND AND CONTROL SYMBOLS - LOCATIONS

We previously discussed point symbols in spreads 25 and 26. Below are some more location point symbols used for command, control and coordination. Expanded notes for symbols may be listed elsewhere such as a legend or table to prevent cluttering the map.

Start Point or other event to occur at a specific location at a specific time, for example: Ammo Supply Pt (ASP), Casualty Coll Pt (CCP), Traffic Control Pt (TCP), Refuel on the Move (ROM), Rearm, Refuel, Resup (R3P), and so on.

Rally point. If the event does not have a specific time, such as an after-action rally, then the symbol may not include the time.

Point of interest. The associated details may not be on the map but contained elsewhere

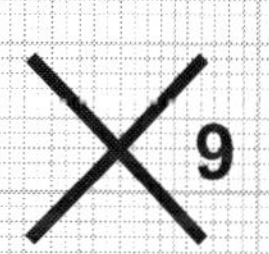

Waypoint. The centre of the crossed lines is the waypoint.

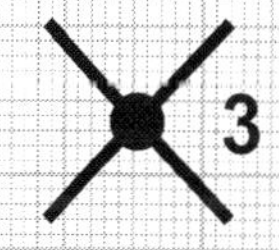

Special Point. The centre of the crossed lines is the point being referred to.

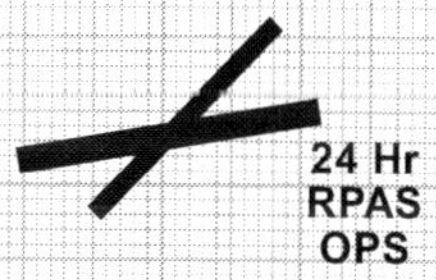

Airfield. The centre of the crossed lines is the location. Free text may be included to provide extra information.

45

PHASE LINES, FEBAS AND FLOTS

Ground manoeuvre symbols represent coordinated activities of friendly forces. They can also represent enemy past and current dispositions and forecast future dispositions.

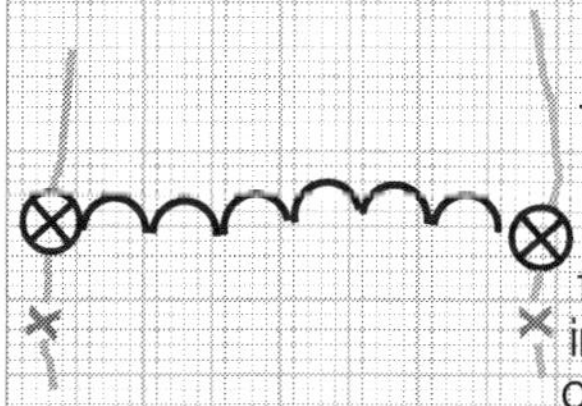

The Forward Line Of Troops (FLOT) is the line indicating the most forward positions of troops in any kind of military operation at a specific time. It is delineated by a series of arcs anchored at two specific co-ordination points. The example shown above is on brigade TAOR boundaries. This symbol can be used to depict enemy FLOTs also, and when friendly and enemy FLOTs are depicted opposite each other, that is the 'Line of Contact'.

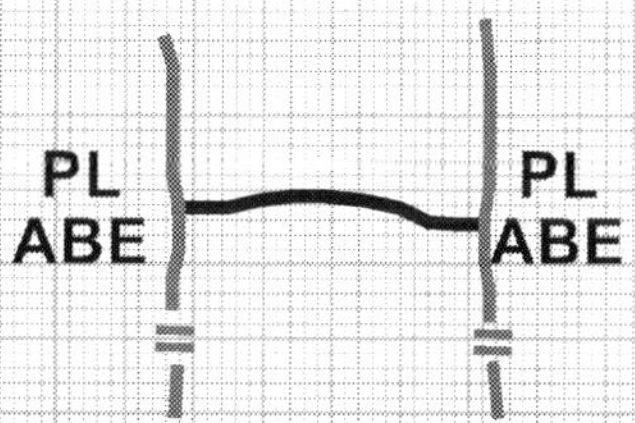

A Phase Line (PL) is a coordination line for an activity. It is often a geographic feature that crosses the AO, or some other means of identifying a line such as an easting or a northing on a map.

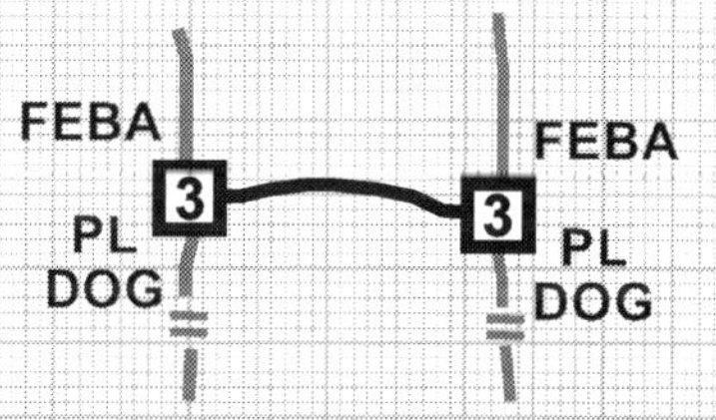

The Forward Edge of the Battle Area (FEBA) is the most forward edge of the areas where ground units are deployed. It may coincide with a phase line.

OFFENSIVE CONTROL MEASURE SYMBOLS

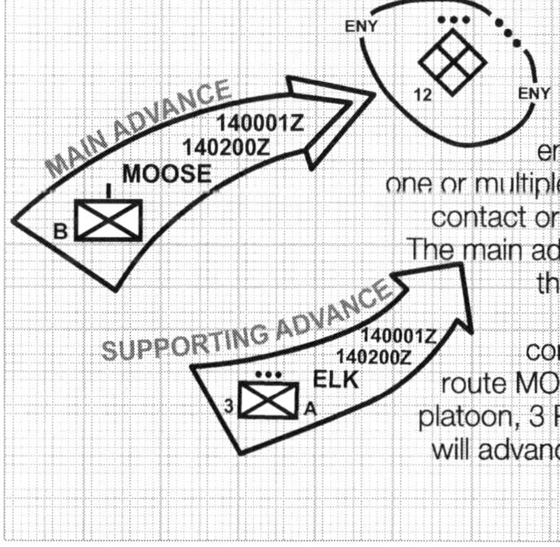

Axis of Advance. A general direction of an advance towards the enemy which may be along one or multiple routes. An advance is in contact or moving towards contact. The main advance will be depicted by the double arrowhead. This diagram shows that B company will advance along route MOOSE while the supporting platoon, 3 Platoon from A Company, will advance at the same time along route ELK.

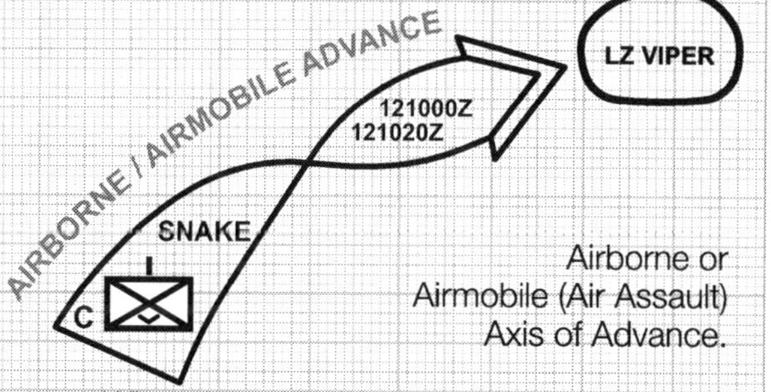

Airborne or Airmobile (Air Assault) Axis of Advance.

When the advance is to be by air, a twisted arrow is used with the appropriate information. Here, C Company (air assault) is being transported for the 20 minute flight along aviation route SNAKE to Landing Zone VIPER.

47

OFFENSIVE CONTROL MEASURE SYMBOLS

Line of attack arrows depict specific routes for an attack. A double arrowhead once again depicts the main attack; a single arrowhead the supporting attack. A dashed arrow is the planned or forecast attack direction effective on order or from a certain time.

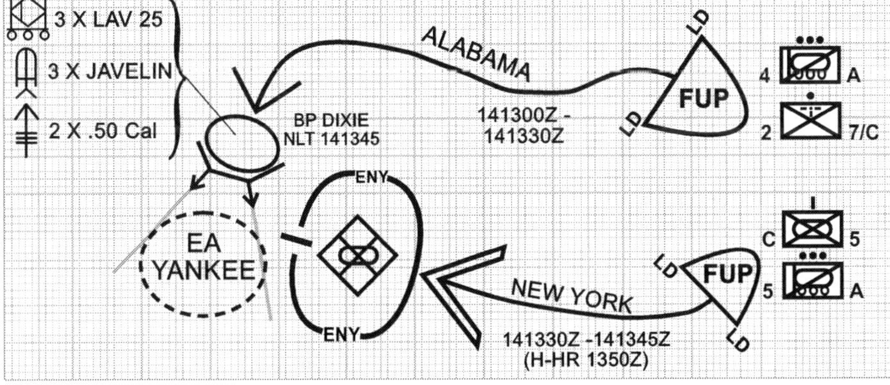

An attack plan using military symbols showing a company of mechanized infantry (C Coy) supported by LAV25s of 5 Troop, A Squadron attacking an enemy mechanized company along attack direction NEW YORK. H Hr 1350. Meanwhile, a supporting force of LAV25s and a section of snipers will travel attack direction ALABAMA, occupy Battle Position Dixie no later than 1345 ready to support by fire and engage reinforcements or withdrawing enemy forces in Engagement Area Yankee with Javelin anti-tank missiles, 25mm guns and .50 cal. sniper rifles. Their tasks from their Form Up Points and crossing the Lines of Departure to the objective are easily inferred. Anyone involved in this attack can get enough information to perform his task.

48

OBSTACLE EFFECTS

Obstacles can be natural or man-made. In warfare, setting the conditions for success includes preparing the battlespace which could include creating obstacles to stop or channel an enemy to the terrain of your choosing.

Basic symbols to make obstacles

Mine (anti tank/anti personnel)

Anti-tank ditch

Anti-tank obstacle

Anti-tank wall

Barbed wire

Concertine wire

Mine field

Anti-tank ditch

Anti-tank wall

Barbed wire (double strand)

Concertine wire (double strand)

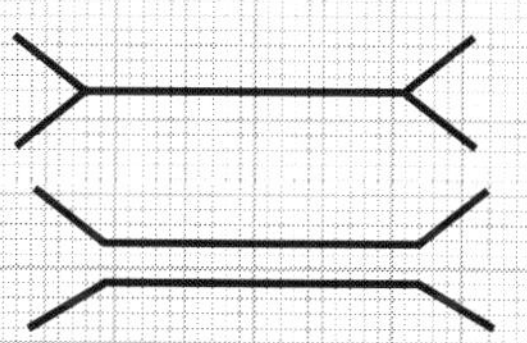

Safe lane through an obstacle

49

TACTICAL MISSION TASKS

Tactical mission tasks, also called 'task verbs', name in one or a few words a specific function or activity to be carried out. Each task has a symbol which can be placed on a map or briefing sketch.

Block - To block the enemy means to deny him access to a given area or to prevent an advance along a route.

The orientation of the 'T' shape will be determined by how the block will be achieved. If achieved by obstacles, terrain and/or fires then the top of the 'T' is the intended limit of enemy advance. If the block is by friendly forces actively opposing enemy forces, then the top of the 'T' faces the enemy.

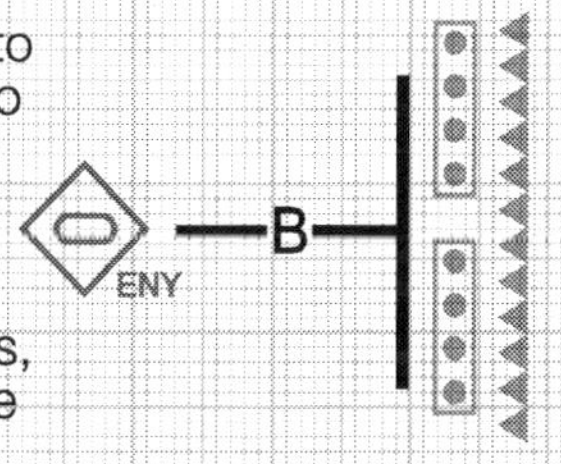

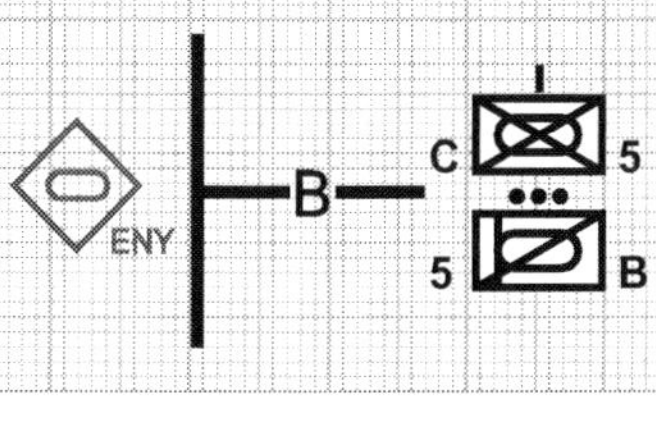

Fix - Similar to block, but to 'Fix' means that enemy forces are held in place by action (arrowhead points toward enemy) or by obstacles (arrow from enemy points towards fixing obstacles).

Fixing the enemy means that he is halted and can be engaged in that location during the time he is fixed.

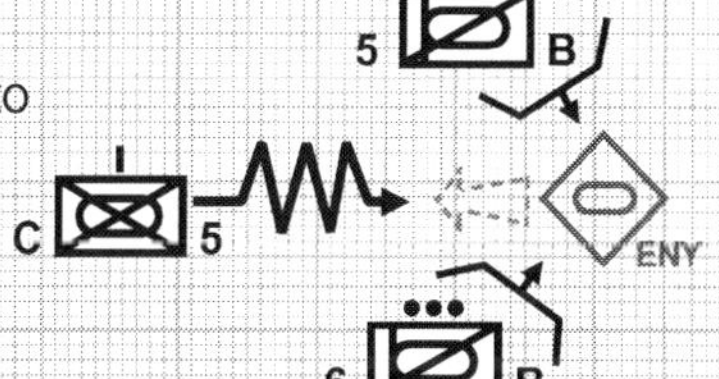

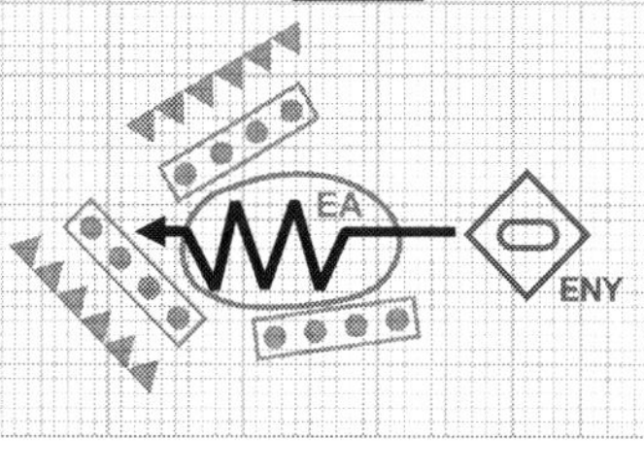

50

TACTICAL MISSION TASKS

Turn - A tactical task or obstacle effect that integrates fire planning and obstacle effort to divert an enemy formation from one avenue of approach to an adjacent avenue of approach or into an engagement area.

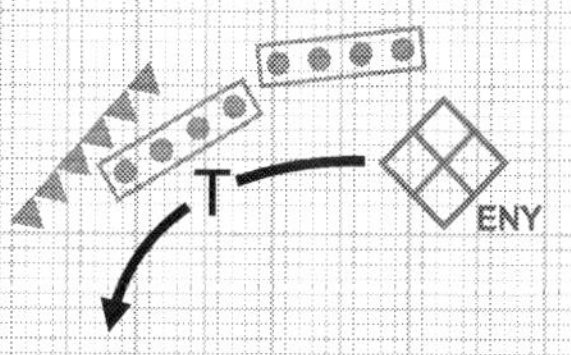

Breach - A tactical task in which the commander employs all available means to break through or establish a passage through enemy defences, obstacles, minefileds or fortifications.

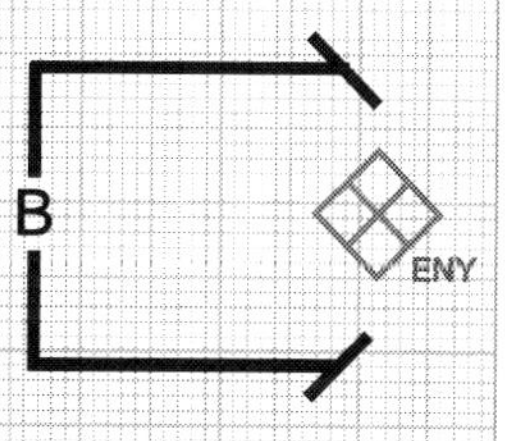

Canalise - A tactical task in which the commander restricts enemy movement to a narrow zone by exploiting terrain coupled with the use of obstacles, fires or friendly manoeuvre.

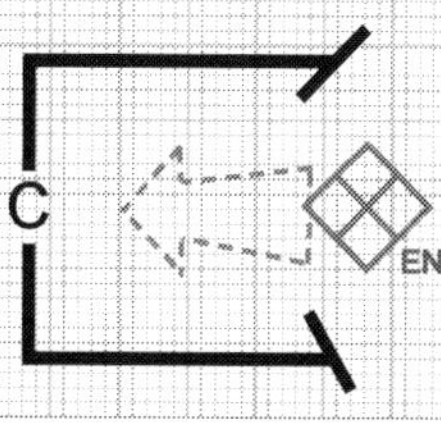

Bypass - To manoeuvre around an obstacle, position, or enemy force to maintain the momentum of the advance while deliberately avoiding combat.

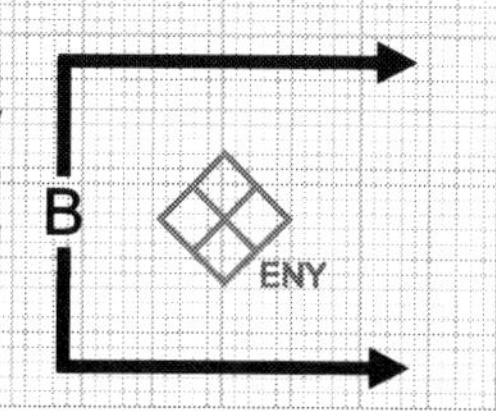

51

TACTICAL MISSION TASKS

Cover - A tactical task to protect the main body by fighting to gain time while also observing and reporting information and preventing enemy ground observation of, and direct fire against, the main body. Unlike a screening or guard force, the covering force is a self-contained force capable of operating independently of the main body.

Guard - A form of security operation whose primary task is to protect the main body while fighting to gain time while also observing and reporting information and preventing enemy ground observation of and direct fire against the main body.

Screen - A form of security operation that provides early warning to the protected force.

The military symbols for Cover, Screen and Guard are the same except for the letter used in the symbol. Here, the symbol shows a squadron of armoured recce conducting a cover. The 'lightning flashes' show the extent of the cover, screen or guard.

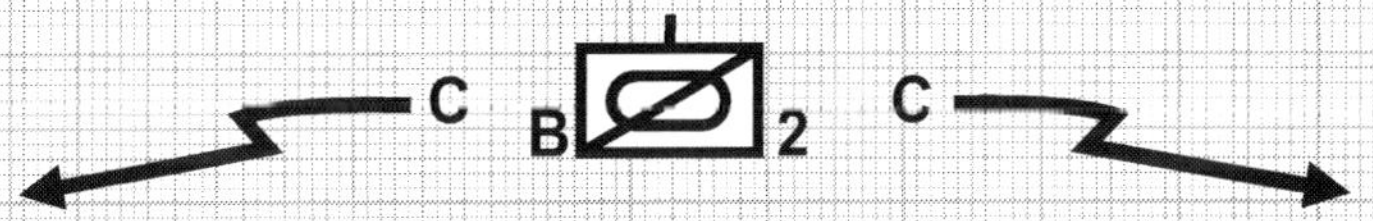

Interdict - A tactical mission where the commander prevents, disrupts or delays the enemy's use of an area or route or capability before it can be used against friendly forces. The arrows should be 45 degrees angled and the centrepoint is the item or area to be targeted.

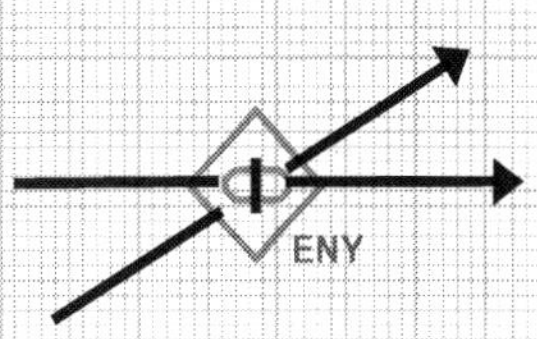

52

TACTICAL MISSION TASKS

Retrograde actions are those that take troops away from the enemy. They may be voluntary and planned such as delay, withdraw or retirement, or they may be forced as a result of enemy action, such as withdraw under pressure.

Delay - A form of retrograde in which a force under pressure trades space for time by slowing down the enemy's momentum and inflicting maximum damage on the enemy without becoming decisively engaged. The symbol includes the planned time to withdraw friendly forces from contact with the enemy. This may be used to allow other friendly forces to withdraw unhindered.

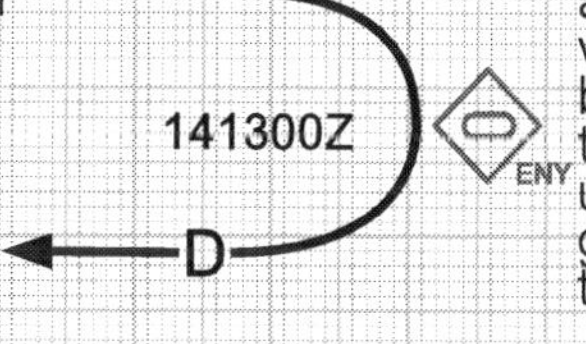

Retirement - A planned form of retrograde in which a force out of contact moves away from the enemy.

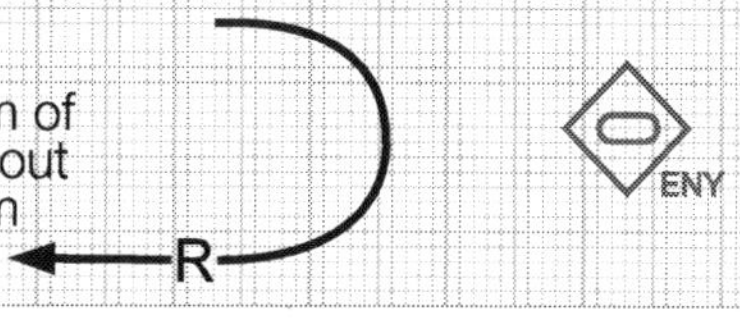

Withdraw - A planned retrograde where a force disengages from the enemy voluntarily. Disengage is similar but may be a quick decision based on unfolding tactical considerations. Both occur usually in order to preserve forces, to gain an advantage in time or location, or to commence another activity.

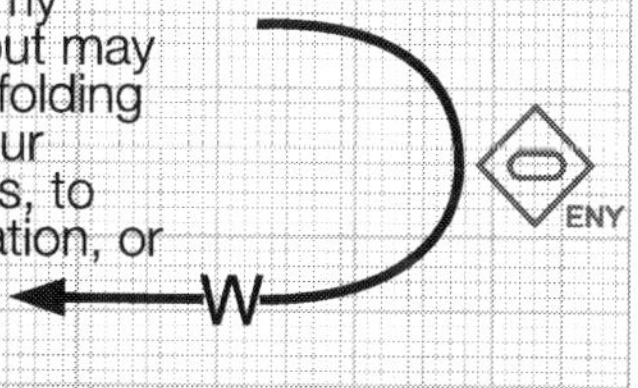

Withdraw under Pressure - A retrograde in which friendly units are forced by enemy pressure to disengage from contact with the enemy, usually in order to preserve forces or to gain a more advantageous position in time or location.

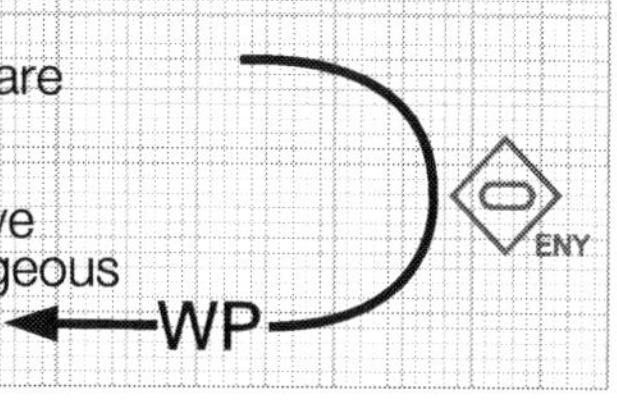

TACTICAL MISSION TASKS CONT.

Because there are many more tactical mission tasks, the following two pages show the most used. Some tasks are US specific and not necessarily common for other nations' militaries, however in a coalition environment it pays to understand them.

Ambush - An attack by fire or other destructive means from concealed positions on a moving or temporarily halted enemy.

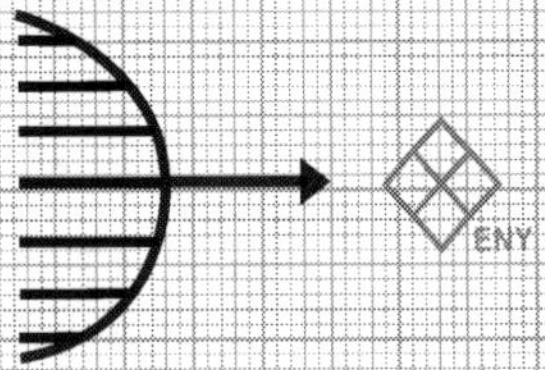

Destroy - To damage a unit or a weapon so seriously that it needs to be rebuilt.

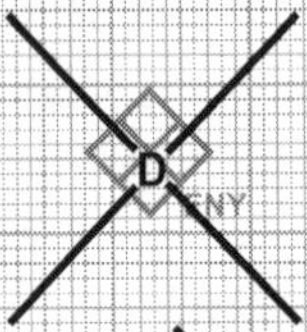

Neutralise - To render enemy personnel or equipment unable to interfere with friendly troops or activities. May or may not involve destroying them.

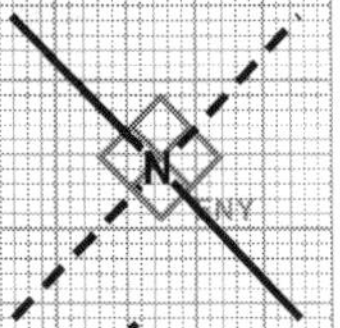

Suppress - To temporarily degrade the ability of personnel or weapons so that they cannot conduct a mission. Depending on the enemy or system, this could be done by electronic jamming, suppressing fire or some other means.

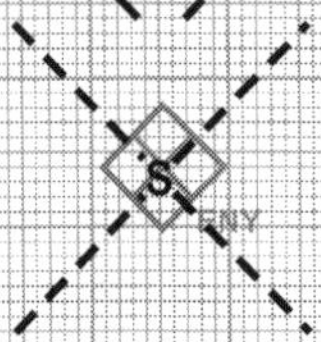

Attack by Fire - A tactical mission task in which a commander uses direct fires, supported by indirect fires, to engage an enemy force without closing with the enemy to destroy, suppress, fix, or deceive that enemy.

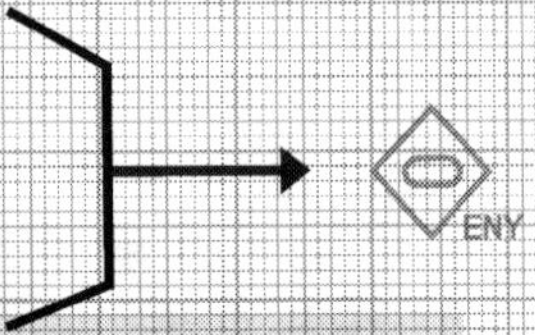

Support by Fire - A tactical mission task in which a manoeuver force moves to a position where it can engage the enemy by direct fire in support of another maneuvering force.

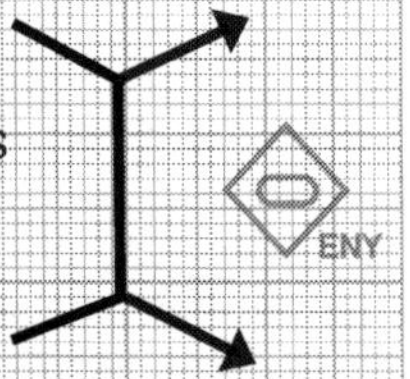

Seize - To use overwhelming force to take possession of a designated area. The symbol or associated graphic includes the unit/s assigned the task.

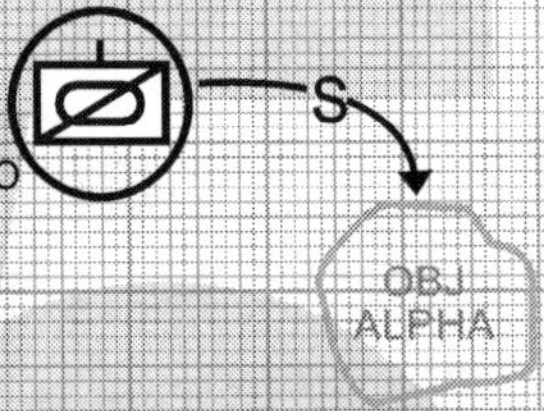

Secure - To prevent a unit, facility or location from being damaged or destroyed as a result of enemy action.

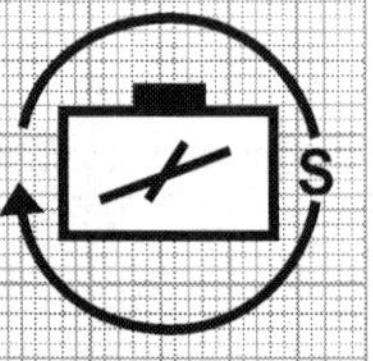

Occupy - To move a force to a designated area so that is under that force's control. Both the movement and occupation occur without enemy opposition.

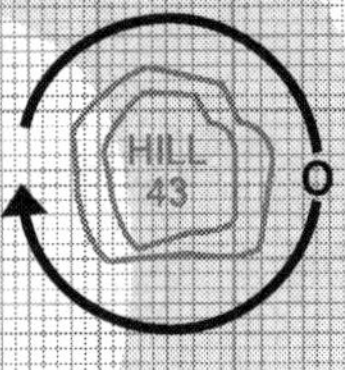

Contain - To stop, hold or surround enemy forces or to cause them to centre their activity on a given front and prevent them from withdrawing any part of their forces for use elsewhere.

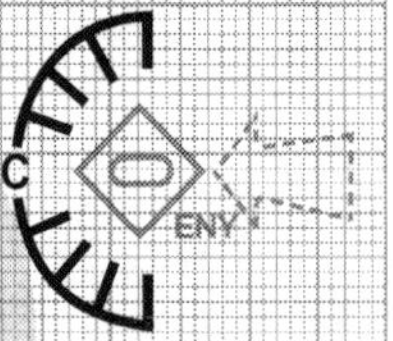

Control - To maintain physical influence over a specified area to prevent its use by the enemy or to create conditions necessary for successful friendly operations. (Note: two arrowheads)

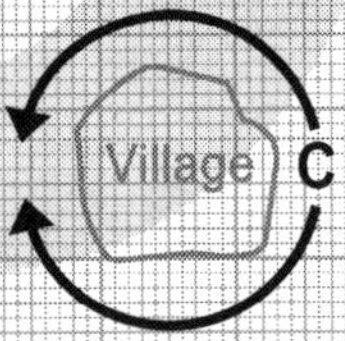

Cordon and Search - A technique of conducting a movement to contact that involves isolating a target area and searching suspect locations within the target area to capture or destroy possible enemy forces and contraband.

Isolate - To seal off, both physically and psychologically, an enemy from sources of support, deny the enemy freedom of movement and to prevent the enemy force from having contact with other enemy forces.

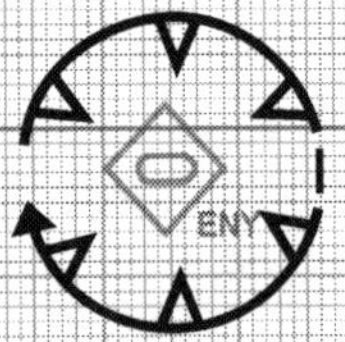

Cordon and Knock - Similar to a Cordon and Search above, but with a level of deference to the householders in the cordon area. Example, military assistance to police during search for civilian suspects of crimes.

Clear - A tactical task that requires the commander to remove all enemy forces and eliminate organised resistance in an assigned area. The symbol may also include a limit of exploitation line.

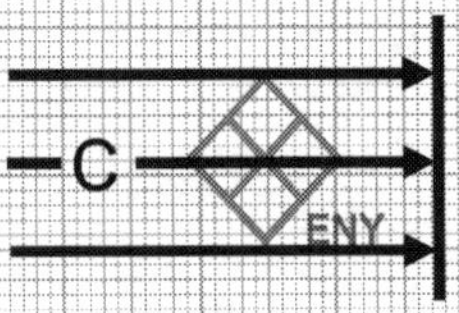

Retain - A tactical mission where the commander ensures that a terrain feature controlled by a friendly force remains free of enemy occupation or use.

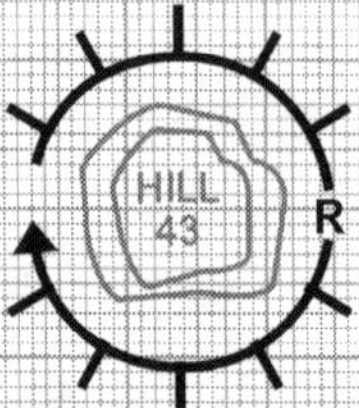

Penetrate - A form of manoeuvre in which an attacking force seeks to rupture enemy defences on a narrow front to disrupt the defensive system.

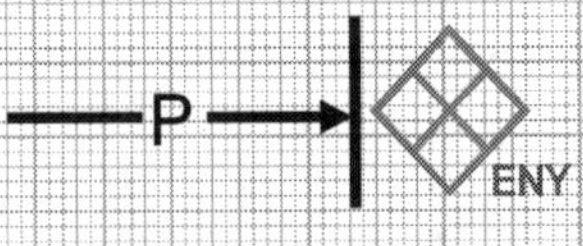

Infiltrate / Exfiltrate - A form of manouevre in which friendly combatants enter enemy territory or depart enemy territory by stealth, surprise or deception without the intention of engaging the enemy during the manoeuvre.

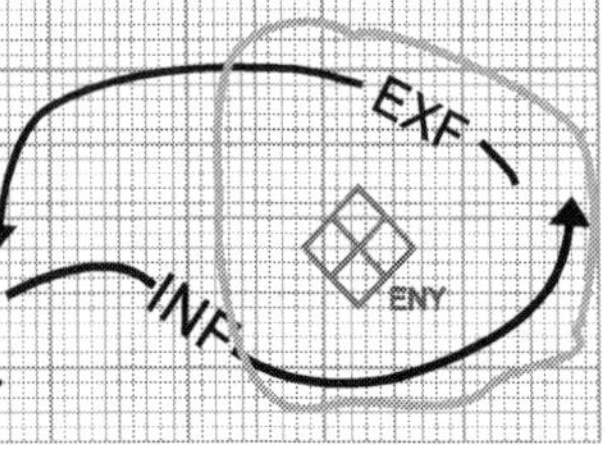

Envelopment - A form of manoeuvre in which an attacking force seeks to avoid the principal enemy defences by seizing objectives behind those defences that allow the targeted enemy force to be destroyed in its current position.

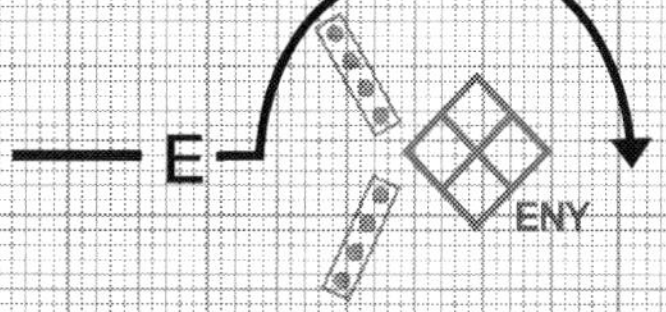

Disrupt - A tactical mission task in which a commander integrates direct and indirect fires, terrain, and obstacles to upset an enemy's formation or tempo, interrupt his timetable, or cause enemy forces to commit prematurely or attack in piecemeal fashion.

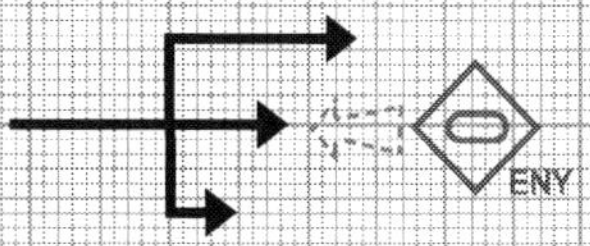

Also, an obstacle effect that focuses fire planning and obstacle effort to cause the enemy to break up his formation and tempo, interrupt his timetable, commit breaching assets prematurely.

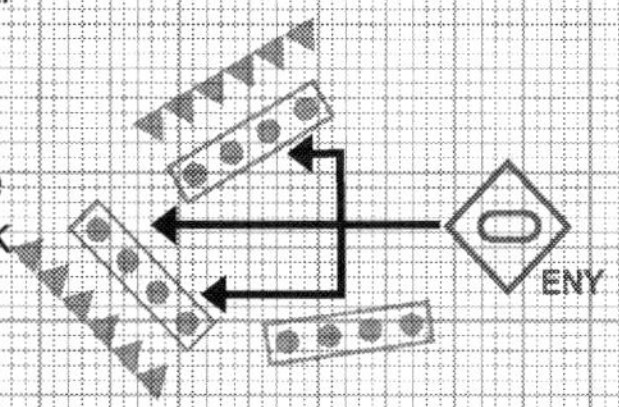

Counterattack - Attack, by part or all of a defending force against an enemy attacking force, for such specific purposes as regaining ground lost, or cutting off or destroying enemy advance units. For sustained counterattacks, the aim is to restore the battle position.
Counterattack by fire - an attack against an advancing enemy by use of stand-off weapons.

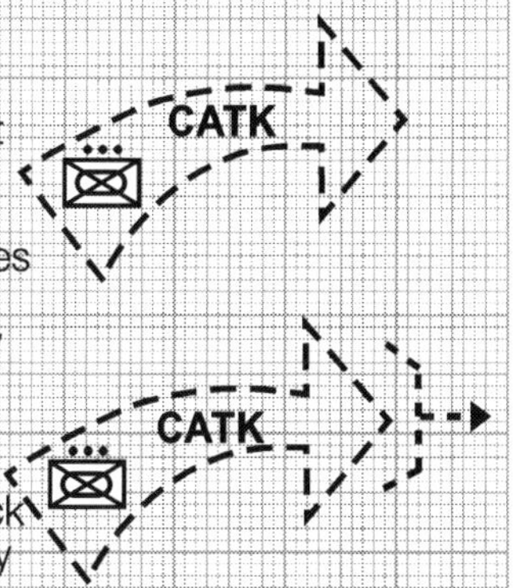

Feint - An offensive action involving contact with the adversary conducted for the purpose of deceiving the adversary as to the location and/or time of the actual main offensive action.

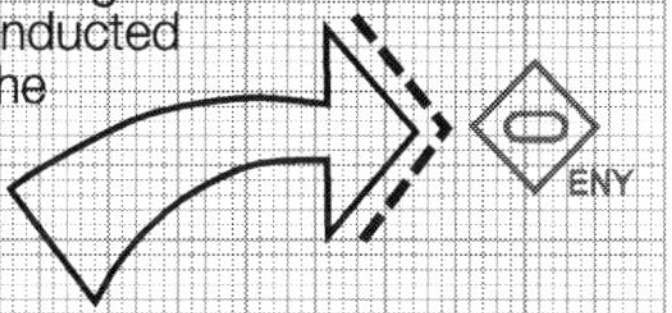

Passage of Lines (Forward) or Passage of Lines (Rearward) - An operation in which a force moves forward or rearward through another force's combat position with the intention of moving into or out of contact with the enemy.

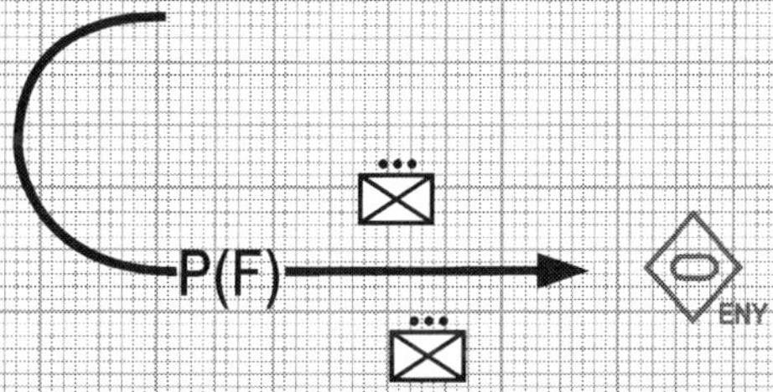

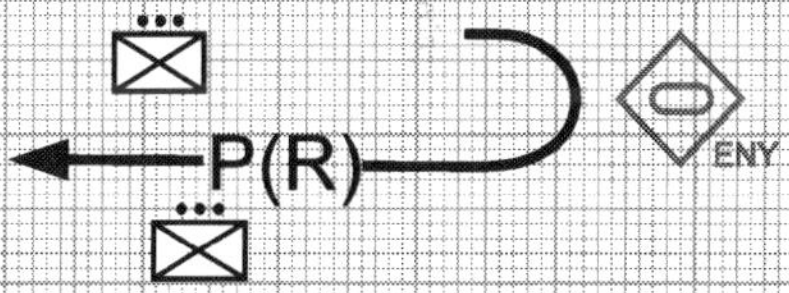

Relief in Place - An operation in which, by direction of higher authority, all or part of a unit is replaced in an area by the incoming unit and the responsibilities of the replaced elements for the mission and the assigned zone of operations are transferred to the incoming unit.

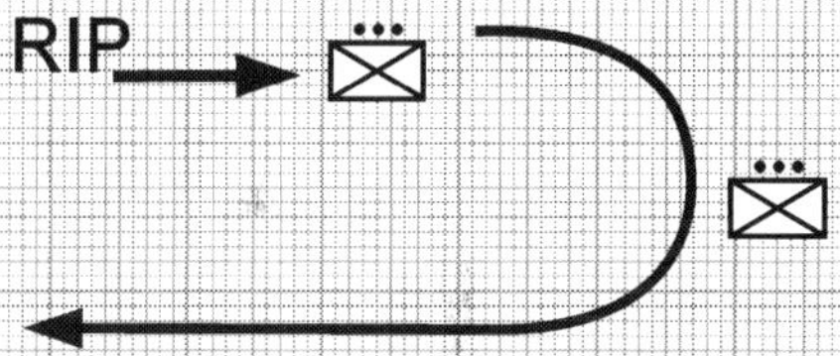

Follow and Assume - A tactical mission task where a second committed force follows a force conducting an offensive task and is prepared to continue the mission if the lead force is fixed, attrited or unable to continue. The follow and assume force is not a reserve but is a depth force committed to specific tasks.

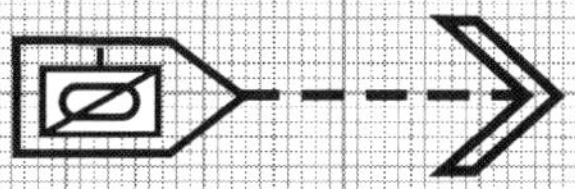

Follow and Support - A tactical mission task where a second committed force follows a force conducting an offensive task and provides support. The follow and support force is not a reserve but is a depth force committed to specific tasks. Note the difference in arrows and arrowheads.

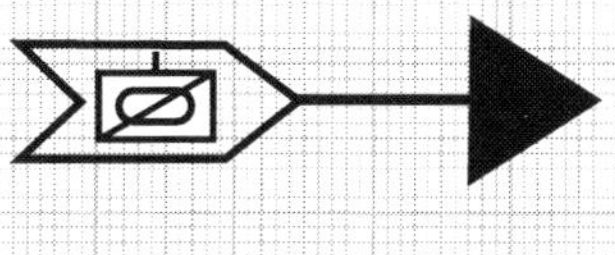

Corrections, amendments or comments?

Feel free to contact the author.

If you have any comments or find any errors in this journal/planner, please contact the author. The information contained herein is a reflection of the current NATO/US doctrine as at time of writing presented in such a way to make it easier to remember. Task definitions are presented as closely as possible to the official glossary but the relevant official manuals remain the authority.
If you have any suggestions for follow up publications, please don't hesistate to contact the author also.

Contact: conway@ipas.com.au
Web: www.ipas.com.au

Made in the USA
Monee, IL
05 November 2026

67385489R00063